Delmar Learning's Test Preparation Series

Automobile Test

Engine Repair (Test A1)

3rd Edition

THOMSON

DELMAR LEARNING

Australia Canada Mexico Singapore Spain United Kingdom United States

THOMSON

TM

DELMAR LEARNING

Delmar Learning's ASE Test Preparation Series

Automobile Test for Engine Repair (Test A1), 3e

Vice President, Technology and Trades SBU:

Alar Elken

Executive Director, Professional Business Unit:

Greg Clayton

Product Development Manager:

Timothy Waters

Developmental Editor:
Christopher Shortt

Channel Manager:
Beth A. Lutz

Marketing Specialist:

Brian McGrath

Production Director:
Mary Ellen Black

Production Manager:

Larry Main

Production Editor:

Elizabeth Hough

Editorial Assistant:
Kristen Shenfield

Cover Images Courtesy of:
DaimlerChrysler Corporation

Cover Designer:

Michael Egan

NOTICE TO THE READER

Contents

Section 1 The History of ASE

Section 2 Take and Pass Every ASE Test

Section 3 Types of Questions on an ASE Exam

Section 4 Overview of the Task List

Section 5 Sample Test for Practice

Section 6 Additional Test Questions for Practice

Section 7 Appendices

Preface

Delmar Learning is very pleased that you have chosen our ASE Test Preparation Series to prepare yourself for the automotive ASE Examination. These guides are available for all of the automotive areas including A1–A8, the L1 Advanced Diagnostic Certification, the P2 Parts Specialist, the C1 Service Consultant and the X1 Undercar Specialist. These guides are designed to introduce you to the Task List for the test you are preparing to take, give you an understanding of what you are expected to be able to do in each task, and take you through sample test questions formatted in the same way the ASE tests are structured. If you have a basic working knowledge of the discipline you are testing for, you will find the Delmar Learning's ASE Test Preparation Series to be an excellent way to understand the "must know" items to pass the test. These books are not textbooks. Their objective is to prepare the technician who has the requisite experience and schooling to challenge ASE testing. It cannot replace the hands-on experience or the theoretical knowledge required by ASE to master vehicle repair technology. If you are unable to understand more than a few of the questions and their explanations in this book, it could be that you require either more shop-floor experience or further study. Some textbooks that can assist you with further study are listed on the rear cover of this book.

Each book begins with an item by item overview of the ASE Task List with explanations of the minimum knowledge you must possess to answer questions related to the task. Following that there are 2 sets of sample questions followed by an answer key to each test and an explanation of the answers to each question. A few of the questions are not strictly ASE format but were included because they help teach a critical concept that will appear on the test. We suggest that you read the complete Task List Overview before taking the first sample test. After taking the first test, score yourself and read the explanation to any questions that you were not sure about, including the questions you answered correctly. Each test question has a reference back to the related task or tasks that it covers. This will help you to go back and read over any area of the task list that you are having trouble with. Once you are satisfied that you have all of your questions answered from the first sample test, take the additional tests and check them. If you pass these tests, you will do well on the ASE test.

Our Commitment to Excellence

The 3rd edition of Delmar Learning's ASE Test Preparation Series has been through a major revision with extensive updates to the ASE's task lists, test questions, and accuracy. Delmar Learning has sought out the best technicians in the country to help with the updating and revision of each of the books in the series.

About the Series Editor

To promote consistency throughout the series, a series advisor took on the task of reading, editing, and helping each of our experts give each book the highest level of accuracy possible. Donny Seyfer has served in the role of Series Advisor for the 3rd edition of the ASE Test Preparation Series. Donny brings to the series several years of experience in writing ASE style questions. Donny is an ASE Master, L1 and C1 certified technician, and service consultant. In 2000 and 2001 Donny received a Regional Technician of the Year award. Donny served as a technical member on several automotive boards. Donny is also the host of an auto care radio show and manages his family repair business in Colorado. Additionally, he revised two of the books in this series and wrote the C1 Service Consultant book.

Thanks for choosing Delmar Learning's ASE Test Preparation Series. All of the writers, editors, Delmar Staff, and myself have worked very hard to make this series second to none. I know you are going to find this book accurate and easy to work with. It is our objective to constantly improve our product at Delmar by responding to feedback. If you have any questions concerning the books in this series, you can email me at: autoexpert@trainingbay.com.

Donny Seyfer
Series Advisor

1 The History of ASE

History

Originally known as The National Institute for Automotive Service Excellence (NIASE), today's ASE was founded in 1972 as a nonprofit, independent entity dedicated to improving the quality of automotive service and repair through the voluntary testing and certification of automotive technicians. Until that time, consumers had no way of distinguishing between competent and incompetent automotive mechanics. In the mid-1960s and early 1970s, efforts were made by several automotive industry affiliated associations to respond to this need. Though the associations were nonprofit, many regarded certification test fees merely as a means of raising additional operating capital. Also, some associations, having a vested interest, produced test scores heavily weighted in the favor of its members.

From these efforts a new independent, nonprofit association, the National Institute for Automotive Service Excellence (NIASE), was established. In early NIASE tests, Mechanic A, Mechanic B type questions were used. Over the years the trend has not changed, but in mid-1984 the term was changed to Technician A, Technician B to better emphasize sophistication of the skills needed to perform successfully in the modern motor vehicle industry. In certain tests the term used is Estimator A/B, Painter A/B, or Parts Specialist A/B. At about that same time, the logo was changed from "The Gear" to "The Blue Seal," and the organization adopted the acronym ASE for Automotive Service Excellence.

ASE

ASE's mission is to improve the quality of vehicle repair and service in the United States through the testing and certification of automotive repair technicians. Prospective candidates register for and take one or more of ASE's many exams.

Upon passing at least one exam and providing proof of two years of related work experience, the technician becomes ASE certified. A technician who passes a series of exams earns ASE Master Technician status. An automobile technician, for example, must pass eight exams for this recognition.

The exams, conducted twice a year at over seven hundred locations around the country, are administered by American College Testing (ACT). They stress real-world diagnostic and repair problems. Though a good knowledge of theory is helpful to the technician in answering many of the questions, there are no questions specifically on theory. Certification is valid for five years. To retain certification, the technician must be retested to renew his or her certificate.

The automotive consumer benefits because ASE certification is a valuable yardstick by which to measure the knowledge and skills of individual technicians, as well as their commitment to their chosen profession. It is also a tribute to the repair facility employing ASE certified technicians. ASE certified technicians are permitted to wear blue and white ASE shoulder insignia, referred to as the "Blue Seal of Excellence," and carry credentials listing their areas of expertise. Often employers display their technicians' credentials in the customer waiting area. Customers look for facilities that display ASE's Blue Seal of Excellence logo on outdoor signs, in the customer waiting area, in the telephone book (Yellow Pages), and in newspaper advertisements.

To become ASE certified, contact:

National Institute for Automotive Service Excellence
101 Blue Seal Drive S.E.
Suite 101
Leesburg, VA 20175
Telephone 703-669-6600
FAX 703-669-6123
www.ase.com

2 Take and Pass Every ASE Test

ASE Testing

Participating in an Automotive Service Excellence (ASE) voluntary certification program gives you a chance to show your customers that you have the "know-how" needed to work on today's modern vehicles. The ASE certification tests allow you to compare your skills and knowledge to the automotive service industry's standards for each specialty area.

If you are the "average" automotive technician taking this test, you are in your mid-thirties and have not attended school for about fifteen years. That means you probably have not taken a test in many years. Some of you, on the other hand, have attended college or taken postsecondary education courses and may be more familiar with taking tests and with test-taking strategies. There is, however, a difference in the ASE test you are preparing to take and the educational tests you may be accustomed to.

Who Writes the Questions?

The questions, written by service industry experts familiar with all aspects of service consulting, are entirely job related. They are designed to test the skills that you need to know to work as a successful technician; theoretical knowledge is not covered.

Each question has its roots in an ASE "item-writing" workshop where service representatives from automobile manufacturers (domestic and import), aftermarket parts and equipment manufacturers, working technicians, and vocational educators meet in a workshop setting to share ideas and translate them into test questions. Each test question written by these experts must survive review by all members of the group. The questions are written to deal with practical application of soft skills and product knowledge experienced by technicians in their day-to-day work.

All questions are pretested and quality-checked on a national sample of technicians. Those questions that meet ASE standards of quality and accuracy are included in the scored sections of the tests; the "rejects" are sent back to the drawing board or discarded altogether.

Each certification test is made up of between forty and eighty multiple-choice questions. The testing sessions are 4 hours and 15 minutes, allowing plenty of time to complete several tests.

Note: Each test could contain additional questions that are included for statistical research purposes only. Your answers to these questions will not affect your score, but since you do not know which ones they are, you should answer all questions in the test. The five-year Recertification Test will cover the same content areas as those listed above. However, the number of questions in each content area of the Recertification Test will be reduced by about one-half.

Objective Tests

A test is called an objective test if the same standards and conditions apply to everyone taking the test and there is only one correct answer to each question. Objective tests primarily measure your ability to recall information. A well-designed objective test can also test your ability to understand, analyze, interpret, and apply your knowledge. Objective tests include true-false, multiple choice, fill in the blank, and matching questions. ASE's tests consist exclusively of four-part multiple-choice objective questions.

Before beginning to take an objective test, quickly look over the test to determine the number of questions, but do not try to read through all of the questions. In an ASE test, there are usually between forty and eighty questions, depending on the subject. Read through each question before marking your answer. Answer the questions in the order they appear on the test. Leave the questions blank that you are not sure of and move on to the next question. You can return to those unanswered questions after you have finished the others. They may be easier to answer at a later time after your mind has had additional time to consider them on a subconscious level. In addition, you might find information in other questions that will help you to answer some of them.

Do not be obsessed by the apparent pattern of responses. For example, do not be influenced by a pattern like **D, C, B, A, D, C, B, A** on an ASE test.

There is also a lot of folk wisdom about taking objective tests. For example, there are those who would advise you to avoid response options that use certain words such as *all, none, always, never, must,* and *only,* to name a few. This, they claim, is because nothing in life is exclusive. They would advise you to choose response options that use words that allow for some exception, such as *sometimes, frequently, rarely, often, usually, seldom,* and *normally.* They would also advise you to avoid the first and last option (A and D) because test writers, they feel, are more comfortable if they put the correct answer in the middle (B and C) of the choices. Another recommendation often offered is to select the option that is either shorter or longer than the other three choices because it is more likely to be correct. Some would advise you to never change an answer since your first intuition is usually correct.

Although there may be a grain of truth in this folk wisdom, ASE test writers try to avoid them and so should you. There are just as many **A** answers as there are **B** answers, just as many **D** answers as **C** answers. As a matter of fact, ASE tries to balance the answers at about 25 percent per choice **A, B, C,** and **D.** There is no intention to use "tricky" words, such as outlined above. Put no credence in the opposing words "sometimes" and "never," for example.

Multiple-choice tests are sometimes challenging because there are often several choices that may seem possible, and it may be difficult to decide on the correct choice. The best strategy, in this case, is to first determine the correct answer before looking at the options. If you see the answer you decided on, you should still examine the options to make sure that none seem more correct than yours. If you do not know or are not sure of the answer, read each option very carefully and try to eliminate those options that you know to be wrong. That way, you can often arrive at the correct choice through a process of elimination.

If you have gone through all of the test and you still do not know the answer to some of the questions, then guess. Yes, guess. You then have at least a 25 percent chance of being correct. If you leave the question blank, you have no chance. In ASE tests, there is no penalty for being wrong.

Preparing for the Exam

The main reason we have included so many sample and practice questions in this guide is, simply, to help you learn what you know and what you don't know. We recommend that you work your way through each question in this book. Before doing this, carefully look through Section 3; it contains a description and explanation of the questions you'll find in an ASE exam.

Once you know what the questions will look like, move to the sample test. After you have answered one of the sample questions (Section 5), read the explanation (Section 7) to the answer for that question. If you don't feel you understand the reasoning for the correct answer, go back and read the overview (Section 4) for the task that is related to that question. If you still don't feel you have a solid understanding of the material, identify a good source of information on the topic, such as a textbook, and do some more studying.

After you have completed the sample test, move to the additional questions (Section 6). This time answer the questions as if you were taking an actual test. Once you have answered all of the questions, grade your results using the answer key in Section 7. For every question that you gave a wrong answer to, study the explanations to the answers and/or the overview of the related task areas.

Here are some basic guidelines to follow while preparing for the exam:

- Focus your studies on those areas you are weak in.
- Be honest with yourself while determining if you understand something.
- Study often but in short periods of time.
- Remove yourself from all distractions while studying.
- Keep in mind the goal of studying is not just to pass the exam, the real goal is to learn!

During the Test

Mark your bubble sheet clearly and accurately. One of the biggest problems an adult faces in test taking, it seems, is in placing an answer in the correct spot on a bubble sheet. Make certain that you mark your answer for, say, question 21, in the space on the bubble sheet designated for the answer for question 21. A correct response in the wrong bubble will probably be wrong. Remember, the answer sheet is machine scored and can only "read" what you have bubbled in. Also, do not bubble in two answers for the same question.

If you finish answering all of the questions on a test ahead of time, go back and review the answers of those questions that you were not sure of. You can often catch careless errors by using the remaining time to review your answers.

At practically every test, some technicians will invariably finish ahead of time and turn their papers in long before the final call. Do not let them distract or intimidate you. Either they knew too little and could not finish the test, or they were very self-confident and thought they knew it all. Perhaps they were trying to impress the proctor or other technicians about how much they know. Often you may hear them later talking about the information they knew all the while but forgot to respond on their answer sheet.

It is not wise to use less than the total amount of time that you are allotted for a test. If there are any doubts, take the time for review. Any product can usually be made better with some additional effort. A test is no exception. It is not necessary to turn in your test paper until you are told to do so.

Your Test Results!

You can gain a better perspective about tests if you know and understand how they are scored. ASE's tests are scored by American College Testing (ACT), a nonpartial, unbiased organization having no vested interest in ASE or in the automotive industry. Each question carries the same weight as any other question. For example, if there are fifty questions, each is worth 2 percent of the total score. The passing grade is 70 percent. That means you must correctly answer thirty-five of the fifty questions to pass the test.

The test results can tell you:

- where your knowledge equals or exceeds that needed for competent performance, or
- where you might need more preparation.

The test results *cannot* tell you:

- how you compare with other technicians, or
- how many questions you answered correctly.

Your ASE test score report will show the number of correct answers you got in each of the content areas. These numbers provide information about your performance in each area of the test. However, because there may be a different number of questions in each area of the test, a high percentage of correct answers in an area with few questions may not offset a low percentage in an area with many questions.

It may be noted that one does not "fail" an ASE test. The technician who does not pass is simply told "More Preparation Needed." Though large differences in percentages may indicate problem areas, it is important to consider how many questions were asked in each area. Since each test evaluates all phases of the work involved in a service specialty, you should be prepared in each area. A low score in one area could keep you from passing an entire test.

There is no such thing as average. You cannot determine your overall test score by adding the percentages given for each task area and dividing by the number of areas. It doesn't work that way because there generally are not the same number of questions in each task area. A task area with twenty questions, for example, counts more toward your total score than a task area with ten questions.

Your test report should give you a good picture of your results and a better understanding of your task areas of strength and weakness.

If you fail to pass the test, you may take it again at any time it is scheduled to be administered. You are the only one who will receive your test score. Test scores will not be given over the telephone by ASE nor will they be released to anyone without your written permission.

3 Types of Questions on an ASE Exam

ASE certification tests are often thought of as being tricky. They may seem to be tricky if you do not completely understand what is being asked. The following examples will help you recognize certain types of ASE questions and avoid common errors.

Each test is made up of forty to eighty multiple-choice questions. Multiple-choice questions are an efficient way to test knowledge. To answer them correctly, you must think about each choice as a possibility, and then choose the one that best answers the question. To do this, read each word of the question carefully. Do not assume you know what the question is about until you have finished reading it.

About 10 percent of the questions on an actual ASE exam will use an illustration. These drawings contain the information needed to correctly answer the question. The illustration must be studied carefully before attempting to answer the question. Often, techs look at the possible answers then try to match up the answers with the drawing. Always do the opposite; match the drawing to the answers. When the illustration is showing an electrical schematic or another system in detail, look over the system and try to figure out how the system works before you look at the question and the possible answers.

Multiple-Choice Questions

The most common type of question used on ASE Tests is the multiple-choice test. This type of question contains 3 "distracters" (wrong answers) and one "key" (correct answer). When the questions are written effort is made to make the distracters plausible to draw an inexperienced technician to one of them. This type of question gives a clear indication of the technician's knowledge. Using multiple criteria including cross-sections by age, race, and other background information, ASE is able to guarantee that a question does not bias for or against any particular group. A question that shows bias toward any particular group is discarded. If you encounter a question that you are unsure of, reverse engineer it by eliminating the items that it cannot be. For example:

A rocker panel is a structural member of which vehicle construction type?

  A. Front-wheel drive
  B. Pickup truck
  C. Unibody
  D. Full-frame

Analysis:

This question asks for a specific answer. By carefully reading the question, you will find that it asks for a construction type that uses the rocker panel as a structural part of the vehicle.

Answer A is wrong. Front-wheel drive is not a vehicle construction type.

Answer B is wrong. A pickup truck is not a type of vehicle construction.

Answer C is correct. Unibody design creates structural integrity by welding parts together, such as the rocker panels, but does not require exterior cosmetic panels installed for full strength.

Answer D is wrong. Full-frame describes a body-over-frame construction type that relies on the frame assembly for structural integrity.

Therefore, the correct answer is C. If the question was read quickly and the words "construction type" were passed over, answer A may have been selected.

EXCEPT Questions

Another type of question used on ASE tests has answers that are all correct except one. The correct answer for this type of question is the answer that is wrong. The word **"EXCEPT"** will always be in capital letters. You must identify which of the choices is the wrong answer. If you read quickly through the question, you may overlook what the question is asking and answer the question with the first correct statement. This will make your answer wrong. An example of this type of question and the analysis is as follows:

All of the following are tools for the analysis of structural damage **EXCEPT:**

A. height gauge
B. tape measure.
C. dial indicator.
D. tram gauge.

Analysis:

The question really requires you to identify the tool that is not used for analyzing structural damage. All tools given in the choices are used for analyzing structural damage except one. This question presents two basic problems for the test-taker who reads through the question too quickly. It may be possible to read over the word **"EXCEPT"** in the question or not think about which type of damage analysis would use answer C. In either case, the correct answer may not be selected. To correctly answer this question, you should know what tools are used for the analysis of structural damage. If you cannot immediately recognize the incorrect tool, you should be able to identify it by analyzing the other choices.

Answer A is wrong. A height gauge *may* be used to analyze structural damage.

Answer B is wrong. A tape measure may be used to analyze structural damage.

Answer C is correct. A dial indicator may be used as a damage analysis tool for moving parts, such as wheels, wheel hubs, and axle shafts, but would not be used to measure structural damage.

Answer D is wrong. A tram gauge *is* used to measure structural damage.

Technician A, Technician B Questions

The type of question that is most popularly associated with an ASE test is the "Technician A says . . . Technician B says . . . Who is right?" type. In this type of question, you must identify the correct statement or statements. To answer this type of question correctly, you must carefully read each technician's statement and judge it on its own merit to determine if the statement is true.

Typically, this type of question begins with a statement about some analysis or repair procedure. This is followed by two statements about the cause of the problem, proper inspection, identification, or repair choices. You are asked whether the first statement, the second statement, both statements, or neither statement is correct. Analyzing this type of question is a little easier than the other types because there are only two ideas to consider although there are still four choices for an answer.

Technician A, Technician B questions are really double true or false questions. The best way to analyze this kind of question is to consider each technician's statement separately. Ask yourself, is A true or false? Is B true or false? Then select your answer from the four choices. An important point to remember is that an ASE Technician A, Technician B question will never have Technician A and B directly disagreeing with each other. That is why you must evaluate each statement independently. An example of this type of question and the analysis of it follows.

Structural dimensions are being measured. Technician A says comparing measurements from one side to the other is enough to determine the damage. Technician B says a tram gauge can be used when a tape measure cannot measure in a straight line from point to point. Who is right?

   A. A only
   B. B only
   C. Both A and B
   D. Neither A nor B

Analysis:

With some vehicles built asymmetrically, side-to-side measurements are not always equal. The manufacturer's specifications need to be verified with a dimension chart before reaching any conclusions about the structural damage.

Answer A is wrong. Technician A's statement is wrong. A tram gauge would provide a point-to-point measurement when a part, such as a strut tower or air cleaner, interrupts a direct line between the points.
Answer B is correct. Technician B is correct. A tram gauge can be used when a tape measure cannot be used to measure in a straight line from point to point.
Answer C is wrong. Since Technician A is not correct, C cannot be the correct answer.
Answer D is wrong. Since Technician B is correct, D cannot be the correct answer.

Most-Likely Questions

Most-Likely questions are somewhat difficult because only one choice is correct while the other three choices are nearly correct. An example of a Most-Likely-cause question is as follows:

The Most-Likely cause of reduced turbocharger boost pressure may be a:

   A. wastegate valve stuck closed.
   B. wastegate valve stuck open.
   C. leaking wastegate diaphragm.
   D. disconnected wastegate linkage.

Analysis:

Answer A is wrong. A wastegate valve stuck closed increases turbocharger boost pressure.
Answer B is correct. A wastegate valve stuck open decreases turbocharger boost pressure.
Answer C is wrong. A leaking wastegate valve diaphragm increases turbocharger boost pressure.
Answer D is wrong. A disconnected wastegate valve linkage will increase turbocharger boost pressure.

LEAST-Likely Questions

Notice that in Most-Likely questions there is no capitalization. This is not so with LEAST-Likely type questions. For this type of question, look for the choice that would be the LEAST-Likely cause of the described situation. Read the entire question carefully before choosing your answer. An example is as follows:

What is the LEAST-Likely cause of a bent pushrod?

  A. Excessive engine speed
  B. A sticking valve
  C. Excessive valve guide clearance
  D. A worn rocker arm stud

Analysis:

Answer A is wrong. Excessive engine speed may cause a bent pushrod.
Answer B is wrong. A sticking valve may cause a bent pushrod.
Answer C is correct. Excessive valve clearance will not generally cause a bent pushrod.
Answer D is wrong. A worn rocker arm stud may cause a bent pushrod.

Summary

There are no four-part multiple-choice ASE questions having "none of the above" or "all of the above" choices. ASE does not use other types of questions, such as fill-in-the-blank, completion, true-false, word-matching, or essay. ASE does not require you to draw diagrams or sketches. If a formula or chart is required to answer a question, it is provided for you. There are no ASE questions that require you to use a pocket calculator.

Testing Time Length

An ASE test session is four hours and fifteen minutes. You may attempt from one to a maximum of four tests in one session. It is recommended, however, that no more than a total of 225 questions be attempted at any test session. This will allow for just over one minute for each question.

Visitors are not permitted at any time. If you wish to leave the test room, for any reason, you must first ask permission. If you finish your test early and wish to leave, you are permitted to do so only during specified dismissal periods.

You should monitor your progress and set an arbitrary limit to how much time you will need for each question. This should be based on the number of questions you are attempting. It is suggested that you wear a watch because some facilities may not have a clock visible to all areas of the room.

4 Overview of the Task List

Engine Repair (Test A1)

The following section includes the task areas and task lists for this test and a written overview of the topics covered in the test.

The task list describes the actual work you should be able to do as a technician that you will be tested on by the ASE. This is your key to the test and you should review this section carefully. We have based our sample test and additional questions upon these tasks, and the overview section will also support your understanding of the task list. ASE advises that the questions on the test may not equal the number of tasks listed.; the task lists tell you what ASE expects you to know how to do and be ready to be tested upon.

At the end of each question in the Sample Test and Additional Test Questions sections, a letter and number will be used as a reference back to this section for additional study. Note the following example: **C.2.**

Task List

C. Engine Block Diagnosis and Repair (18 Questions)

Task C.2 **Visually inspect engine block for cracks, corrosion, passage condition, expansion and gallery plug holes, and surface warpage; determine necessary action.**

Example:

29. Technician A says a warped cylinder head mounting surface on an engine block may cause valve seat distortion during assembly. Technician B says a warped cylinder head mounting surface on an engine block may cause coolant and combustion leaks. Who is right?
   A. A only
   B. B only
   C. Both A and B
   D. Neither A nor B   (C.2)

Question #29
Answer A is wrong.
Answer B is wrong.
Answer C is correct. A warped cylinder head mounting surface on the cylinder block will cause a cylinder head to bend as it is bolted down. This may cause the valve seats to distort. It may also allow coolant and combustion gases to leak past the head gasket.
Answer D is wrong.

Task List and Overview

A. General Engine Diagnosis (17 Questions)

Task A.1   Verify driver's complaint and/or road test vehicle; determine necessary action.

In the modern automotive repair shop environment, a customer service specialist or Service Consultant is usually responsible for collecting information from the customer to assist the technician in resolving a customer's concern. Armed with this information, his skills, and a fair amount of intuition, the technician must determine the cause of the particular concern.

One of the most important tools a technician can utilize is the pre-test drive. During this drive he may be able to determine the problem and other conditions that may contribute to or relate to the concern. After verifying the customer's concern the technician must determine the most efficient diagnostic path to take. Most experienced technicians will check with available information systems to see if the problem they are diagnosing is described in a technical service bulletin or service campaign.

Once the problem is found and resolved a final test drive allows the technician to confirm that the problem has indeed been resolved and that no other problems have come to light after the repair.

Task A.2   Determine if no-crank, no-start, or hard starting condition is an ignition system, cranking system, fuel system, or engine mechanical problem.

If the starter fails to crank the engine, the problem may range from a faulty starter motor to broken components inside the engine. If no sounds come from the starter motor when it is activated, first disable the ignition system. Then attempt to rotate the crankshaft pulley by hand in the normal direction of rotation. If the crankshaft can be rotated freely through two complete revolutions, then diagnosis of the vehicles starting system is the next step.

If you are unable to rotate the crankshaft by hand, the engine may be hydrostatically locked or have broken internal components. To check for hydrostatic lock, remove all the spark plugs and attempt to rotate the crankshaft again. If oil or coolant squirts from the spark plug holes, this indicates a bad head gasket, warped cylinder head or block, or a cracked cylinder head or block.

If the crankshaft cannot be rotated at all with the spark plugs removed, or cannot be rotated through at least one complete revolution, the engine may be seized or have broken internal parts. Pull the dipstick and check crankcase oil level. If oil does not register on the dipstick, it is possible that the pistons are seized in their bores or the connecting rods are seized to the crankshaft. If the oil level is sufficient, a broken component may have lodged between moving parts inside the cylinder block, preventing the parts from rotating.

Many overhead camshaft (OHC) engines are non-freewheeling or "interference" engines. On these engines, a no-crank condition may be caused by piston-to-valve contact. This is a common occurrence when a timing belt slips or breaks, but it may also occur on engines fitted with a timing chain and sprockets. On many belt-driven OHC engines it is possible to easily loosen or remove part of the timing belt cover. Do this, if possible, and check for obvious signs of belt failure.

If the customer states that the starter cranks the engine but it will not start (or takes a long time to start), confirm that the valve train is operating properly before attempting to crank the engine yourself. If the timing belt or chain is broken or jumped, additional cranking may cause severe engine damage.

A no-start or hard starting complaint can be caused by a faulty ignition, fuel, or emission control system. These complaints can also be caused by broken or slipped valve train timing components, especially on free-wheeling engines. A broken timing device may cause some cylinders to have good compression while others have none. A slipped timing device may result in all cylinders having low compression. To determine if the belt or chain is functioning properly, rotate the crankshaft by hand while observing the distributor rotor or camshaft. If these components fail to rotate with the crankshaft, the timing belt or chain is broken. If they do rotate with the crankshaft, confirm proper indexing of the rotor or camshaft to determine if the belt or chain has slipped. Rotate the crankshaft until the piston in cylinder #1 is at TDC on the compression stroke. Then check distributor rotor or camshaft position to make sure that it is correct.

In the real world most technicians will check for the presence of spark before checking for fuel or mechanical issues. The process for checking spark requires using a tool called a "spark tester" in place of one of the spark plugs and cranking the engine to visually check for spark. Some distributorless ignition systems are not compatible with this type of test and may be tested using appropriate secondary ignition lab scope procedures.

Fuel injected engines usually receive high-pressure fuel from an electric pump. To verify that the fuel pump is operating and fuel is reaching the engine, locate the fuel line that supplies fuel to the throttle body or fuel injector rail. Turn the key on/engine off and install a fuel pressure gauge. Verify that the gauge registers adequate pressure. If the gauge does not register any pressure or registers very low pressure, proceed with fuel system diagnosis. If fuel pressure is adequate, begin diagnosis of the fuel injection control system.

Task A.3   Inspect engine assembly for fuel, oil, coolant, and other leaks; determine necessary action.

The source of fluid leaks can be difficult to locate. Determining what type of fluid is leaking will reduce the number of possible leak locations.

Engine oil usually leaks from faulty gaskets and seals, but it can also leak from cracked castings, faulty pressure switches or sending units, and loose tapered oil gallery plugs. Oil can leak from an area high on the engine (like a V-type engine intake manifold rear seal) and run down the engine, appearing at the rear of the oil pan. Do not assume that the "wet" area is the source of the leak. Clean the area and run the engine to check for fresh fluid. Also, do not immediately assume that a leaking seal or gasket is faulty. Excessive blowby or a faulty PCV system can pressurize the crankcase, forcing oil past a seal or gasket that is in good condition. Hard-to-find oil leaks can be located by pouring a small quantity of fluorescent dye into the crankcase and running the engine. When an ultraviolet light is shined onto the engine, oil containing the dye will glow to reveal the leak point.

Fuel may leak from loose connections or damaged components. Check for loose hose clamps and fuel line fittings. Check hoses for swelling, cracks, and damage from abrasion. Check metal lines for cracks and corrosion. Check for leaking O-ring connections and a leaking fuel pressure regulator (often mounted on the throttle body unit or injector rail).

The soft metal plugs that seal the cooling system channels in the cylinder block and sometimes the cylinder heads are known by many names. We will use expansion plug in this text but be aware that they are also known as expansion or freeze plugs and that these terms are interchangeable.

Corroded expansion plugs are common coolant leakage points, as are faulty hoses and water pumps. Check coolant temperature sensors, sending units, and thermal vacuum switches, too. Some engines have expansion plugs at the back of the cylinder block and/or head. If the engine consumes coolant, but you cannot find evidence of leaking coolant, check the engine oil level and condition. Coolant may be leaking into the crankcase. Coolant can also leak into the combustion chambers.

On vehicles equipped with power steering, check the fluid level. Leaking power steering fluid may be mistaken for engine oil or transmission fluid.

Task A.4   Listen to engine noises; determine necessary action.

Different types of engine part failures often make distinctive sounds. First, be sure that the noise is actually coming from the engine. A faulty water pump, alternator, power steering pump, A/C compressor or air injection pump can make noises that appear to be coming from inside the engine. Loose or broken accessory mounting brackets can also cause noises that sound like engine internal problems. Listen to each of the accessories using a stethoscope to determine if it is the source of a noise. If in doubt, temporarily remove the drive belt from an accessory to prevent it from operating.

A faulty crankshaft main or rod bearing usually makes a knocking sound that is very deep in pitch. Main bearing knock is usually a thumping noise most noticeable when the engine is first started. Connecting rod bearings also cause a heavy knocking sound, and engine oil pressure may also be low, especially at idle. When the faulty cylinder is disabled during a cylinder balance test, the knocking sound will diminish. Loose flywheel bolts may cause a thumping noise at idle. Camshaft bearings usually do not cause a noise unless severely worn.

Worn pistons and cylinders cause a rapping noise while accelerating. When performing a cylinder balance test, piston noise can increase when the faulty cylinder is disabled (the opposite reaction of a bad connecting rod bearing). A piston pin with excessive clearance often makes a "double click" noise when the engine is idling.

Lifters also make a distinctive noise, a loud ticking sound. One way to isolate lifter (or other valve train) noise from connecting rod noise is to remember that the camshaft operates at half of crankshaft speed. It is common for a lifter with excessive leak-down to tick for a few seconds after the engine starts. The noise goes away once full oil pressure is developed.

Task A.5   Diagnose the cause of excessive oil consumption, coolant consumption, unusual engine exhaust color, odor, and sound; determine necessary action.

Excessive oil consumption can be due to oil leaking from the engine or oil being drawn into the cylinders and burned. Before blaming internal components, be absolutely sure that oil is not leaking from the engine. In some cases oil leaks only when the engine is running. If necessary, raise the vehicle on a lift while the engine is running to check for leaks.

Oil can enter the cylinders several different ways, including: worn rings; scored cylinder walls; worn valve guides, seals, and stems; worn turbocharger seals; and plugged oil drain passages. As a general rule, an engine that is "burning oil" will emit blue-gray exhaust. This may be more noticeable on acceleration and deceleration. Do not confuse the blue-gray smoke due to oil consumption with the black exhaust that occurs when the air/fuel ratio is too rich.

Plugged oil drain passages in the cylinder head or block can cause excessive oil consumption even when rings and guides are in good condition. To check for this, remove the oil filler cap or another component fitted to a valve cover and start the engine. If the oil level inside the cover rises steadily as the engine runs and reaches the top of the valve guides, the drain passages are clogged. While these passages can usually be cleared of sludge, the sludge is an indication that the engine was poorly maintained. Clearing the passages will probably reduce oil consumption, but the engine may experience other problems in the near future.

Perform compression tests (Task A.8) and cylinder leakage tests (Task A.9) to confirm that piston rings/cylinders or valve guides are worn.

If the engine is turbocharged, first perform oil consumption diagnosis as though the engine was not turbocharged. While turbos are commonly blamed for excessive oil consumption problems, about half of the turbos returned under warranty are not defective. If oil is found in the turbo compressor housing or intake manifold, check the oil drain from the turbo housing to the block. If it is obstructed, oil under pressure will be forced into the engine. Check the PCV system, too. If the PCV valve does not close during "boost" conditions, the crankcase will be pressurized. This may pressurize the turbo oil drain passage, forcing oil into the turbo housing.

Like oil consumption, coolant consumption may be caused by coolant leaking from the cooling system or coolant leaking into the engine (or passenger compartment). First, eliminate external leaks as the cause of coolant consumption by performing cooling system pressure tests (Task D.3). If cooling system pressure drops during the tests, but no leaks are found, check the engine oil level and condition. Leaks into the crankcase will raise the oil level. If coolant is being drawn into the combustion chambers, the exhaust will be gray or white. The engine will continue to emit this smoke long after the time it usually takes for moisture to be purged from the exhaust system. On vehicles equipped with electronic engine controls, coolant passing through the exhaust system will "poison" the oxygen sensor.

Some engine problems can be diagnosed by listening to the exhaust pulses at the tailpipe. If all cylinders are firing properly, the exhaust should consist of steady pulses. A puffing noise that occurs at regular intervals usually indicates a cylinder misfire caused by a compression, ignition, or fuel system defect. Puffing noises that occur erratically are usually caused by ignition or fuel system defects. Engine idle speed may also be unsteady.

A high-pitched squealing noise during hard acceleration may be caused by a small leak in the exhaust system, particularly in the exhaust manifolds or exhaust pipe. The leak may also be noticeable at idle as a ticking noise.

Another common engine noise is a high-pitched whistle at idle and low engine speeds. Check for vacuum leaks at the intake manifold gaskets. Also check for cracked or disconnected vacuum hoses. A vacuum leak whistle gradually decreases when the engine is accelerated and the intake vacuum decreases.

A strong sulfur, or rotten egg, smell coming from the exhaust system of a car fitted with a catalytic converter may indicate a rich air/fuel ratio.

Task A.6   Perform engine vacuum tests; determine necessary action.

A vacuum test can be used to help pinpoint the cause of an engine problem. The vacuum gauge should be connected directly to the intake manifold.

On an engine that is performing correctly, the vacuum gauge reading should be between 17 and 22 in. Hg (45 and 28 kPa absolute) and steady with the engine idling. Some abnormal vacuum gauge readings and typical problems associated with them are listed below.

- A slightly low but steady reading may indicate late ignition timing.
- A very low but steady reading may indicate that the intake manifold has a significant leak.
- Burned or leaking valves cause the vacuum gauge to fluctuate.
- Weak valve springs may result in a vacuum gauge fluctuation.
- A leaking head gasket may cause a vacuum gauge fluctuation.
- If the valves are sticking, the vacuum gauge fluctuates.
- If, when the engine is accelerated and held steady at a higher speed, the vacuum gauge pointer gradually falls, the catalytic converter or other exhaust system components are restricted.

Task A.7   Perform cylinder power balance tests; determine necessary action.

In a sound engine each cylinder contributes equal amounts of power. A common method of isolating a power loss that may be occurring in part of the cylinders is the power balance test. In most engines, this test is performed by disabling spark on a single cylinder for a short time to measure the RPM drop that occurs when that cylinder is not contributing. Cylinders that are not contributing or are weak will show little or no RPM drop when they are disabled. On port fuel injected engines many manufacturers include the ability to perform cylinder balance tests with the PCM and a scanner. Most OBD II vehicles will quickly pick-up on and identify an inefficient cylinder, displaying a code directing to the perceived problem and the affected cylinder.

When a particular cylinder or cylinders fail a power balance test, it becomes necessary to determine if the cause is mechanical; valves, internal engine efficiency, or engine management related. Engine management issues could include port fuel injectors that leak or are electrically damaged (delivering no fuel at all), faulty ignition components like spark plugs, DIS coils, plug wires, etc. The power balance test is usually the precursor to a cylinder compression and or leak down test. These tests will confirm a mechanical issue and electrical diagnosis using digital storage oscilloscopes or scanners can determine engine management issues.

Task A.8   Perform cylinder compression tests; determine necessary action.

The ignition and fuel injection system must be disabled before proceeding with the compression test. During the compression test, the throttle is blocked open and the engine is cranked through four compression strokes for each cylinder. The compression readings are recorded for each stroke and compared to the manufacturer's specifications. Slightly low compression readings in all cylinders are not cause for concern if engine performance is acceptable. Compression readings that vary more than 20 percent (from the highest to the lowest) are cause for concern. Compression readings may be interpreted as follows:

- When the compression readings on all the cylinders are about equal, but significantly lower than specifications, the piston rings or cylinder walls may be worn. If compression in all cylinders is low and the engine spins freely during cranking, check the valve timing. The timing belt or sprocket may have jumped.

- Low compression readings on one or more cylinders indicates worn rings, leaking valves, a blown head gasket, flat camshaft, or a cracked cylinder head. Performing a leak down test will narrow down the cause of the problem.

- Low compression readings in two adjacent cylinders is probably due to a leaking head gasket or cracked cylinder head.

- Zero compression in a cylinder is usually caused by a hole in a piston or a severely burned exhaust valve. If the zero compression reading is caused by a hole in the piston, the engine will have excessive blowby.

- Higher than specified compression usually indicates carbon deposits in the combustion chamber.

Task A.9   Perform cylinder leakage tests; determine necessary action.

During a cylinder leakage test, a regulated amount of air from the shop air supply is forced into the cylinder while both the exhaust and intake valves are closed. The gauge on the leakage tester indicates the percentage of leakage in the cylinder. A gauge reading of 0 percent indicates that there is no cylinder leakage. If the reading is 100 percent, the cylinder is not holding any air.

If cylinder leakage exceeds 20 percent, determine the source of the leakage by listening for air escaping through the tailpipe, crankcase (via the oil filler cap or PCV valve), and or the intake tract (throttle body or carburetor). Air escaping from the tailpipe indicates an exhaust valve leak. When the air is coming out of the PCV valve or valve cover opening, the piston rings are leaking. An intake valve is leaking if air is escaping from the top of the throttle body or carburetor. Remove the radiator cap and check the coolant for bubbles, which indicates a leaking head gasket or cracked head.

B. Cylinder Head and Valve Train Diagnosis and Repair (14 Questions)

Task B.1   **Remove cylinder heads, disassemble, clean, and prepare for inspection according to manufacturer's procedures.**

Remove a cylinder head only when the engine is cold. Removing a warm cylinder head may cause the head to warp, especially if it is made of aluminum.

Remove the cylinder head bolts, loosening the bolts in a sequence *opposite* that of the tightening sequence. Note and record the positions of special bolts. Remove the cylinder head from the engine. Cylinder heads can be quite heavy, so ask an assistant to help you, especially if the engine is still mounted in the vehicle.

Use a spring compressor to compress the valve springs and then remove the valve locks, or "keepers." Release the compressor and remove the retainer, rotator, spring, and spring seats from the head. Keep all parts in an organizer so they can be returned to their original cylinder. Check the valve stem tips for mushrooming. If it is present, the tip must be dressed with a file before the valves are removed from the head. Remove the valves from the cylinder head and place them in an organizer.

When removing the cylinder head from an overhead camshaft (OHC) design engine, the timing belt or chain must first be disconnected from the camshaft. The procedure for doing this varies from manufacturer to manufacturer. On some engines with a chain-driven camshaft, the camshaft sprocket is unbolted from the cam. The cylinder head assembly is then removed, leaving the chain and sprockets in position on the engine. On some engines with a belt-driven camshaft, the belt tensioner is loosened and the belt is slipped off the camshaft sprocket. On other engines the timing cover and belt must be completely removed from the engine. If the timing belt is being removed from the engine and will be reused, mark the direction of rotation on the belt. Reinstall the belt so it rotates in the same direction. Never crank the engine after a timing device has been loosened or removed (until the cylinder head has been removed). Cranking an engine while the timing belt is loose or disconnected can cause immediate and serious engine damage.

The basic cylinder head removal procedure varies from manufacturer to manufacturer. On some OHC engines the cylinder head, camshaft(s), and rocker arms (if used) are removed as an assembly after loosening and removing the cylinder head bolts. On some engines the rocker shaft and arm assembly (including the upper half of the cam bearings) must first be removed to access cylinder head mounting bolts. Refer to the appropriate service manual for information.

Cleaning the components for reassembly changed dramatically with the advent of aluminum heads and blocks. Many of the surfaces of these components are polished to near mirror-like qualities. After cleaning these components in the appropriate degreasing solution, carbon build-up may be removed by hand with a soft wire wheel or in an abrasive cabinet like a glass blaster. In either case, care must be taken to insure that gasket surfaces remain in their original condition. Use of refinishing wheels on polished surfaces can cause new gaskets to fail prematurely. The head must be carefully washed after removing deposits and old gasket material to insure that none of the debris restricts passages. Many manufacturers are recommending plastic scrapers and gasket softening chemicals to remove old gaskets. Be sure to find out what the manufacturer recommends to guarantee a quality repair.

Task B.2   **Visually inspect cylinder heads for cracks and gasket surface areas for warpage, corrosion, and leakage; check passage condition; determine necessary repair.**

While the cylinder heads from any engine should be carefully inspected, the heads from engines with serious mechanical problems (blown head gasket, coolant consumption, overheating, oil sludging, etc.) should receive special attention. First look at the old head gaskets to determine if a problem area is visible. If one (or more) is, match the area to the contact area on the cylinder head.

Check the cylinder head for cracks, paying special attention to the combustion chambers and the areas between intake and exhaust valves. An electromagnetic-type tester and iron filings (MAGNA-FLUX®) may be used to check for cracks in cast-iron heads. A dye penetrant may be used to locate cracks in aluminum heads. Machine shops can usually locate hard-to-find cracks by pressure testing. In this type of test, all coolant passages are blocked using metal plates. The coolant jacket is then filled with compressed air and the head is submerged in a tank of water. Bubbles escaping from the head reveal leak areas.

Use a straightedge and a feeler gauge to check the cylinder head for warpage at several locations. Check the manufacturer's service manual for exact specifications. A cylinder head that is excessively warped must be resurfaced or replaced.

On overhead camshaft (OHC) engines, be sure to check warpage on the cam side of the head before resurfacing and reinstalling the head. If warpage exceeds manufacturer's specs, the camshaft will bind, flex, and may break.

Inspect the coolant passages in the cylinder head as thoroughly as possible. Shine a flashlight into the passages, looking for corrosion, rust, and trapped debris. A cylinder head that shows evidence of severe pitting in the cooling jacket should be replaced.

Task B.3   Inspect and test valve springs for squareness, pressure, and free height comparison; replace as necessary.

A valve spring tester is used to measure valve spring pressure or tension. Valve springs must be carefully checked to see that the free height of the spring meets specifications and that the spring does not lean to one side. The spring must stand exactly perpendicular when it sits on a flat surface. The spring must be able to attain a specified pressure when at rest. This is called the installed height. It must also be able to attain specific pressure at maximum valve lift. A spring tester is used to measure these pressures at their respective heights. In most cases, if a spring fails it will not be able to develop adequate pressure. This can cause the valves to "float" or bounce when the engine is at higher revs. Low spring pressure often leads to a spring breaking and the valve falling into the cylinder. In the more unusual event that the spring generates too much pressure, it can cause excessive wear on the cam, followers, or lifters.

Task B.4   Inspect valve spring retainers, rotators, locks, and valve lock grooves.

Valve spring retainers and locks must be checked for wear, scoring, or damage. When any of these conditions are present, replace the components.

Valve rotators can be inspected before cylinder head removal and disassembly since they usually cannot be taken apart. Rotators can be located on top of the valve spring (built into the spring retainer) or between the valve spring and the cylinder head. To test a rotator, mark the top of the spring retainer with a dab of paint. Then start the engine and run it at about 1,500 rpm. The retainer should slowly rotate. The direction of rotation is not important. If the retainer does not rotate, replace the rotator.

In some cases, a defective rotator can be diagnosed with the cylinder head removed. Check the valve stem tip wear pattern. A shallow groove or channel running across the tip indicates that the valve has not been rotating. Replace the rotator on any valve with this condition.

Inspect the valve lock grooves machined into the valve stems. Look for damage and wear, particularly for round shoulders. If the shoulders are uneven or rounded, replace the valve. Valve lock failures cause severe engine damage.

Task B.5   Replace valve stem seals.

It is not necessary to remove the cylinder head to service many valve train components. This includes the valve springs, oil seals, retainers, and valve locks. A claw type valve spring compressor can be used to compress the valve spring while the head is mounted on the block.

On engines equipped with umbrella type valve seals, slip the new seals over the valve stems. Work carefully to avoid damaging new seals on valve stem lock grooves. A damaged seal will cause excessive oil consumption. Some seals come with an installation tool. The tool is simply a short plastic sleeve that is slipped over the tip of each valve stem before the seal is installed. The sleeve extends down far enough to cover the lock grooves. After installing each seal, push it down against the top of the valve guide and remove the installation tool.

On engines equipped with positive type valve seals, the installation procedure is the same with one important difference. Positive type seals must be pushed down over the top of the valve guide. Each seal has some sort of retaining device to keep it attached to the guide. Some seals have a flat, circular spring that wraps around the seal. Some use garter springs to hold the seal in place. Others have a molded-in ridge on their inner diameter that mates with a groove machined onto the valve guide. Whatever the retaining method, be sure that the valve guide seal is securely attached to the guide. Some manufacturers specify that a special tool should be used to drive the positive seal onto the valve guide. For further information, refer to a service manual for the vehicle.

Task B.6   Inspect valve guides for wear; check valve guide height and stem-to-guide clearance; determine needed repairs.

Valve guides should be measured near the top, center, and bottom using a hole gauge. Measure the valve stem diameter with a micrometer in the same three positions, and subtract the stem readings from the guide measurements to obtain the clearance. An alternate method for measuring stem-to-guide clearance is to install the valve in the guide with the valve about ⅛ inch (3.18 mm) off its seat. Mount a dial indicator against the valve margin or against the valve stem below the lock grooves. Move the valve from side to side while observing the clearance reading on the dial indicator. Divide the reading by two to obtain stem-to-guide clearance.

If clearance exceeds specifications, the valve guides may be replaced or bored out, and a thin-wall liner installed. Excessive valve stem-to-guide clearance may result in an improper valve seating and lower compression. Increased oil consumption may result from excessive valve stem-to-guide clearance.

Valve guide height is usually measured from the top of the spring seat to the top of the guide. If guide height is not within specifications, check to see if a pressed-in guide was improperly installed or has moved in the cylinder head.

Task B.7   Inspect valves and valve seats; determine needed repairs.

Although most shops no longer perform valve and seat reconditioning, technicians still need to be able to inspect the condition of valves and seats to determine the cause of failure. Older engines and many large displacement light truck engines are more susceptible to wear. The valves in these engines are larger than small engines so they tend to have more trouble holding their shape and dissipating heat into the valve seats. In these larger valves it is not uncommon to see the exhaust valves tulip or crack under hard use. Here again, aluminum heads eliminated most of these conditions due to their ability to draw excessive heat out of the exhaust valves. The downside is that the conditions that used to cause valve damage will now, more likely, damage the cylinder head instead. Engines with multiple valves and smaller displacement engines demonstrate excellent valve and seat life. In most cases the valves and seats will last the life of the engine. Common failures are: valve damage when timing belts break in interference engines and carbon build-up on the valve tulip that causes poor contact with the seat and loss of compression.

When inspecting valves and seats we are looking for obvious changes in the shape of the valve. The seat to valve contact area is critical to good sealing. The valve train is designed to lift the valve and gently set it down. If excessive valve lash occurs in adjustable valve trains or due to a failure of a hydraulic lifter or lash adjuster that can cause accelerated valve seat wear. Any irregularity in this surface can cause a valve to not seal resulting in cylinder misfire. The valves must seal completely for the cylinder to reach full pressure during both the compression and power strokes.

Task B.8   Check valve face-to-seat contact and valve seat concentricity (runout).

After grinding the valves and resurfacing the valve seats, check valve face-to-seat contact. Apply blue dye (Prussian blue) to the valve seat and install the valve against the seat. Tap the valve head, but *do not* rotate the valve. Remove the valve and observe the dye on the valve face to determine the width and height of the transfer area (contact area). The width and height of the contact area can be changed by revising the topping or throating angles in the valve seat.

The transfer area also shows whether or not the valve and seat are concentric. If the blue dye transfers 360° around the valve face, the valve and seat are concentric. If the blue dye does not appear 360° around the valve face, replace the valve.

A valve seat concentricity tester containing a dial indicator may also be used to measure valve seat concentricity.

Task B.9   Check valve spring installed (assembled) height and valve stem height; determine needed repairs.

Measure the installed valve stem height from the spring seat surface on the cylinder head to the valve stem tip. If stem installed height is greater than specifications, the valve stem is stretched or too much material has been removed from the valve face or seat. Install a new valve and measure stem height again. If the measurement is still excessive, replace the seat or cylinder head. Excessive valve stem height moves the plunger downward in a hydraulic valve lifter and may cause valve train components to bottom out.

Measure the installed valve spring height from the lower edge of the top retainer to the spring seat. If this measurement is excessive, install shims between the bottom of the valve spring and the top of the spring seat surface on the cylinder head. Excessive valve spring installed height reduces valve spring tension, which may result in valve float and cylinder misfiring at higher speeds.

Task B.10   Inspect pushrods, rocker arms, rocker arm pivots, and shafts for wear, bending, cracks, looseness, and blocked oil passages; repair or replace as required.

Pushrods should be inspected for a bent condition and wear on the ends. Roll the pushrod on a level surface to check for a bent condition. Bent pushrods usually indicate interference in the valve train, such as a sticking valve, improper valve adjustment, or mechanical interference due to improper valve timing. If the pushrod has an oil passage to provide oil to the rocker arm, make sure that the passage is not obstructed.

Worn rocker arms, shafts, or pivots cause improper valve adjustment and a clicking noise in the valve train. Check rocker arm shafts for wear and scoring in the rocker arm contact area. Check the shafts for cracks, bending, and loose/leaking oil passage plugs (if fitted). Check rocker arms for scoring at the pivot area and valve stem tip contact area. Worn rocker arms should be replaced.

Task B.11   Inspect and replace hydraulic or mechanical lifters/lash adjusters.

Technicians usually diagnose hydraulic valve train problems from the outside in. In most cases the customer is complaining of a ticking or clicking noise in the engine. We use our stethoscope to pinpoint the location of the noise and perform the necessary teardown to arrive at the problem component. It is important to understand how the hydraulic lash adjuster, this includes the traditional V8 type lifter, functions. These components are built to amazingly small tolerances often .0001 of an inch. They have a spring in them that holds the valve train components in place. This spring is not responsible for controlling valve lash however, that is a function of engine oil pressure.

The hydraulic lash adjuster maintains a preload on valve train components at all times when the engine is running.

Hydraulic lash adjusters range in size from about one inch in diameter in cam-in-block engines to as small as .25 of an inch in many of the smaller multi-valve engines. Due to their very close tolerances, they are not very compatible with dirty oil, which is the main cause for failure. In applications where the lifter (lash adjuster) rides directly on the camshaft, there are 2 designs; the flat tappet and the roller design.

The flat tappet design starts life with a slightly convex bottom that encourages it to spin as the angled face of the camshaft raises and lowers it. This motion keeps oil moving over the cam and lifter contact surfaces. When inspecting this type for damage, look for dull surfaces, pitting, or concave contact surfaces. If this kind of damage is present, the camshaft will have to be replaced as well.

The roller lifter was originally used in racing engines because of its ability to follow large and complex camshaft lobe profiles. Due to the fact that a cylinder rolls on the camshaft surface the contact area between the cam and lifter is very small and has significantly less friction than the flat tappet design. Original equipment manufacturers began using this design to reduce internal friction and net better gas mileage. The hydraulic portion of this design is the same as all the others. The component that rolls will generally not fail unless it fails to roll. Here again watch for signs of pitting or for components that do not roll easily. The roller dictates that the lifter cannot spin so the cam and roller are made of steel. In most applications a used roller lifter that has not failed can be installed on a new camshaft. This is not possible with a flat tappet design where new lifters may be installed on an old cam but used lifters cannot be installed on a new cam.

When a hydraulic failure occurs in any of the 3 designs the unit is replaced with a new component. In some rare instances the lash adjuster can fail in such a way that it holds the valve open. This can cause valve damage particularly on exhaust valves.

Task B.12 Adjust valves on engines with mechanical or hydraulic lifters.

Valve lash adjusting procedures and mechanisms vary from manufacturer to manufacturer. On some engines valves are adjusted while the engine is cold. On others, the engine should be at operating temperature. Refer to the appropriate service manual for instructions.

On some engines with mechanical valve lifters, the rocker arms have an adjustment screw and a locknut on the valve stem end of the arm. Other engines use an interference fit screw without a locknut. The adjustment procedure usually involves rotating the crankshaft to position the piston in the cylinder being adjusted at top dead center (TDC) on the compression stroke. Feeler gauges are then inserted between the adjusting screw and the valve stem. If clearance is excessive, the adjustment screw locknut (if equipped) is loosened and the adjustment screw is turned. When clearance is correct, a feeler gauge of the correct thickness will slide between the adjusting screw and the valve stem with a light push fit. Tighten the locknut (if equipped) when clearance is correct.

Some OHC engines fitted with mechanical valve lifters have removable metal pads in each lifter or spring retainer. Clearance is measured by placing feeler gauges between the cam lobe and the lifter or spring retainer while the piston is at TDC on the compression stroke. Pads are available in various thicknesses to adjust the clearance to specifications.

Some valve trains have hydraulic valve lifters and individual rocker arm pivots retained with self-locking nuts. These valve trains require an initial adjustment of the rocker arm nut to position the lifter plunger. With the valve closed, loosen the rocker arm nut until there is clearance between the end of the rocker arm and the valve stem. Slowly turn the rocker arm nut clockwise until a zero lash condition exists. This occurs when the lifter is not preloaded but all of the components have just come in contact with each other. The next step is to apply the preload specified by the manufacturer. In most applications this is the equivalent of .030–.060 of an inch of lifter pre-load. It is important to follow the manufacturers specs here as the lifter must be able to compensate for expansion and contraction of valve train components during cold to fully warm engine operation.

The valve train on engines with hydraulic lifters and stud-mounted rocker arms can also be adjusted while the engine is running. After removing the valve covers, install oil shrouds on the rocker arms to prevent oil from splashing onto the exhaust manifolds and other nearby parts. Start the engine and loosen the rocker retaining nut until a clicking noise begins. Then slowly tighten the nut just until the clicking noise stops. From this point, slowly tighten the nut about ¼ turn at a time the specified number of turns. Wait a few seconds between each ¼ turn to allow the lifter to leak down. Turning the adjusting too much at one time or too far can cause piston-to-valve contact.

Task B.13 Inspect and replace camshaft (includes checking gear wear and backlash, sprocket and chain wear, overhead cam drive sprockets, drive belts, belt tension, tensioners, and cam sensor components).

When the camshaft gear teeth mesh directly with the crankshaft gear teeth, gear backlash may be measured with a dial indicator positioned against one of the camshaft gear teeth. Rock the cam gear back and forth and note the maximum reading on the indicator.

Some engines equipped with a timing chain and sprockets are fitted with a hydraulic tensioner. The tensioner uses pressurized oil from the lubrication system to eliminate timing chain play. Some manufacturers recommend measuring the installed length of the tensioner to determine chain wear. If the tensioner length exceeds the manufacturer's specifications, replace the timing chain.

On many V-type camshafts in block engines, a timing mark on the crankshaft sprocket must be aligned with a timing mark on the camshaft sprocket before the camshaft sprocket and chain are installed. Timing chain stretch and wear may be measured on these engines with a socket and flex handle installed on one of the camshaft sprocket retaining bolts. Rock the camshaft sprocket back and forth without moving the crankshaft gear and measure the movement on one of the chain link pins on the camshaft sprocket.

Engines fitted with a timing belt should have the belt replaced at the mileage intervals recommended by the engine manufacturer. During belt replacement, the crankshaft, camshaft, and idler or other sprockets should be inspected. Check the sprocket teeth for wear and damage. Check idler pulleys or sprockets for dry or loose bearings.

Many vehicles with distributor-less ignition systems (DIS) substitute a camshaft position sensor for the distributor to find the location of the #1 cylinder. This is most often used to synchronize fuel injection events to ignition events. The cam sensor is usually a magnetic hall-effect switch that gets a signal from the camshaft via a #1 aligned bump or machined surface on the cam or cam drive gear.

When inspecting camshafts, look for pitting, dull surfaces or indications of poor lubrication. Cam or crank gears on chain drives develop wear that appears like the tooth of the chain in reverse. This condition warrants replacement. Chain deflection is the primary indicator of timing chain condition. Refer to the manufacturers' recommendations to determine chain condition. Many chain driven engines use chain guides that are tensioned by the hydraulic chain tensioner. Be sure to check for surface wear on these. If the chain has worn a groove in the guide, it will probably require replacement. Reassembly of timing components in engines with hydraulic tensioners may require that the tensioner be compressed and pinned until all components are in place.

Task B.14 Inspect and measure camshaft journals and lobes; measure camshaft lift.

To check camshaft straightness, rest the camshaft outer bearing journals on V-blocks and position a dial indicator against the middle cam bearing journal. Rotate the camshaft to determine runout. If the camshaft is not straight, replace it.

To measure camshaft lobes, use a micrometer to measure from the highest point on the lobe to a point on the opposite side of the lobe. Record the measurement. Then measure the lobe again at a position 90° from the first measurement. Subtracting the second measurement from the first gives camshaft lobe lift. Replace the camshaft if lift is not within specifications.

Use a micrometer to measure the diameter of each camshaft journal. Measure this diameter in several locations. If the diameter is less than specified, replace the camshaft.

Camshaft lobe lift can also be measured with the camshaft still in the engine by removing the valve cover and mounting a dial indicator on the cylinder head. Position the dial indicator so that it contacts the pushrod tip (rocker arm removed) or rocker arm directly above the pushrod (rocker arm still in place). The dial indicator stem must be parallel with the pushrod. Crank the engine by hand and note the highest and lowest dial indicator readings. The difference between these two numbers is camshaft lobe lift.

Task B.15 Inspect and measure camshaft bore for wear, damage, out-of-round, and alignment; determine needed repairs.

Inspect the camshaft bearing bores on a camshaft-in-block for scoring or other damage. Minor nicks or burrs can be sanded. If the bearing bores are severely damaged, the block should be replaced.

On overhead cam engines without removable bearing caps, check the bearing bores for damage. Minor nicks or burrs can be sanded. If the camshaft is binding, or the bearing inserts are worn unevenly, use a straightedge to check bearing bore alignment. If the bores are out of alignment, the head is warped and should be straightened or replaced.

On overhead cam engines with removable camshaft bearing caps, the camshaft usually runs directly against the aluminum head. After removing the bearing caps and the cam, place a straightedge across the cam bearing surfaces to measure bearing alignment. Measure the clearance between the straightedge and each bearing bore to determine the bore alignment. When the camshaft bearing bores are improperly aligned, replace the cylinder head. Check the bearing surfaces for scoring and other damage. To measure bearing bore out-of-round, install the bearing caps and torque the retaining bolts to specifications. Then measure bore diameter at several locations around the bore using a telescoping gauge. If the bearing surfaces and bores are in good condition, use Plastigage to measure the bearing clearance.

Task B.16 Time camshaft(s) to crankshaft.

With the timing belt or chain cover removed, camshaft timing may be checked by noting the positions of marks on the camshaft and crankshaft sprockets. These marks must be aligned as indicated in the vehicle manufacturer's service manual.

On many OHV pushrod engines, the crankshaft sprocket is installed on the crankshaft nose and the crankshaft is rotated to position piston #1 at TDC. At this point, a mark stamped onto the crankshaft sprocket is pointing directly upward (toward the camshaft). The camshaft sprocket is then temporarily bolted to the cam and used to rotate the cam until a mark stamped on the cam sprocket is pointing directly downward (toward the crankshaft). The sprocket is then removed from the cam (without allowing the cam to rotate). The timing chain is looped over the cam gear, the mark on the cam gear is positioned directly downward, and the chain is looped around the crankshaft sprocket. When the cam sprocket is attached to the cam, the timing marks on the crank and cam sprockets should be pointing toward one another.

Single overhead camshaft engines fitted with a timing belt often use a similar procedure. After positioning the crankshaft so that piston #1 is at TDC, the camshaft is rotated to align a mark on the cam sprocket with a mark on the cylinder head. The timing belt is then installed.

The procedure used to time camshafts on double overhead camshaft engines varies from manufacturer to manufacturer. On some engines the cam sprockets are friction fitted to the cams. On these engines the cams can be rotated after the timing belt is installed. When the cams are rotated to the proper positions, the bolts locking the cam sprockets to the cams are tightened. Other DOHC engines use a procedure similar to that of many SOHC engines.

Valve timing may be checked by observing the valve position in relation to the piston position. With any piston at top dead center (TDC) on the compression stroke, the intake and exhaust valves for that cylinder should be completely closed. When the piston is at TDC on the exhaust stroke, the intake valve should be opening, and the exhaust valve should be closing. This position is called valve overlap. If the valves do not open properly in relation to the crankshaft position, the valve timing is not right. Incorrect valve timing may cause low power or, in extreme cases, bent valves due to piston-to-valve contact.

Task B.17 Inspect cylinder head mating surface condition and finish, reassemble and install gasket(s) and cylinder head(s); replace and tighten fasteners according to manufacturers' procedures.

Clean and inspect the cylinder block deck in preparation for head installation. Make sure that all head positioning dowels, if used, are in place in the block. Run a tap into cylinder head bolt threaded holes. Then use compressed air to eject any debris from the threaded holes. Always wear eye protection when using compressed air to clean surfaces or openings. Allowing debris or fluid to remain in threaded holes will cause false torque readings when the head bolts are tightened. Coolant or combustion leaks may result. If the holes are blind holes, fluid or debris at the bottom of the holes may cause the block to crack when the bolts are tightened.

Many newer engines are fitted with torque-to-yield (TTY) cylinder head bolts. These bolts are usually tightened to a specific torque and then rotated tighter a specified number of degrees. Torque-to-yield bolts are permanently stretched as they are tightened and produce a more uniform clamping force. Most, but not all, TTY bolts must be replaced with new bolts once they are loosened. Check the manufacturer's service manual for information.

Most modern head gaskets are installed dry, without any type of sealer. When positioning a head gasket on the block, make sure that any orientation marks (up, front, left, right, etc.) are followed.

After double-checking the cylinder bores for tools, shop towels, dropped fasteners, etc., set the cylinder head on the engine block. Check the manufacturer's recommendation regarding thread lubricants or sealers. Bolts that are threaded into blind holes are often lubricated with a few drops of engine oil-some on the threads and some on the underside of the bolt head. Bolts that thread into the water jacket are often coated with a waterproof sealer.

Insert the head bolts in their holes and hand tighten them. Then tighten the bolts to specifications following the procedure and sequence specified by the engine manufacturer.

C. Engine Block Diagnosis and Repair (14 Questions)

Task C.1 Disassemble engine block and clean and prepare components for inspection.

Mount the engine on a stand and remove the oil pan drain plug. Allow any oil that has accumulated in the pan during engine removal to drain into a pan. Remove the lifters from the block and store them in an organizer for later inspection. Turn the engine block upside down, remove the oil pan bolts, and remove the pan from the block. If the pan is "glued" to the block with RTV sealant, strike a strong corner of the pan with a rubber mallet to loosen it.

Check the crankshaft and connecting rod bearing caps to see if they are marked for position and direction. Main bearing caps often have numbers and arrows cast into each cap. Arrows typically point to the front (timing device end) of the engine. Connecting rods are often stamped with the cylinder number on both the rod body and cap, near the cap parting line. Take note of the direction that the numbers (or rod oil squirt holes) point for all connecting rods. If rod or main caps are not marked, mark each one with a number punch, center punch, or scratch awl.

Inspect the top of the cylinder bores for ring ridges. If the ring ridge is severe, it should be removed *before* attempting to remove the piston/connecting rod assemblies. Use a ridge reamer to remove the ridge (see Task C.4).

Loosen the connecting rod cap bolts or nuts and remove the caps. Keep used bearing inserts with their caps for later inspection. If bolts are held captive in the connecting rod bodies, place a short length of fuel line hose over each bolt to protect the crankshaft journals during piston/rod removal.

Carefully push each piston/connecting rod assembly out of its cylinder. When an assembly is removed, immediately reinstall its mating rod cap and nuts.

Remove the harmonic balancer or pulley hub bolt from the crankshaft nose and remove the balancer or hub. Some simply slide off the nose, but most are a press fit. Use a special harmonic balancer removal/installation tool to remove a press-fit balancer. Using a jawed puller to remove a press-fit balancer will permanently damage the balancer.

Remove the timing chain/belt cover bolts and remove the cover. On engines fitted with timing chain tensioning devices, compress and lock the tensioner shoe in place, if possible. Remove the tensioner and any timing chain guides from the front of the block. Remove the oil slinger, if one is present, from the crankshaft nose.

On camshaft-in-block engines, unbolt and remove a bolted-on camshaft sprocket along with the timing chain. Remove the camshaft thrust plate bolts and the thrust plate, if one is present. If the engine has a balance shaft mounted in the engine "V," remove the shaft drive mechanism.

On engines fitted with a one piece crankshaft rear main oil seal, pry the seal out of its bore or unbolt the seal mounting plate and seal from the back of the engine. Remove the crankshaft main bearing cap bolts and the bearing caps. Lift the crankshaft out of the block.

On cam-in-block engines, carefully withdraw the camshaft from the cylinder block, being careful to avoid nicking the bearing bores or the camshaft lobes. Use a camshaft bearing installer/remover tool to drive the bearings from their bores.

If the engine has block-mounted balance shafts, remove them now, following the engine manufacturer's instructions.

Knock any expansion plugs loose by striking one edge with a blunt chisel. Do not strike the plug too hard and do not try to drive the plug straight into the water jacket. This could cause a bulge in a cylinder wall if the plug is driven into the wall. When the expansion plug tilts in its bore, grab an edge of the plug with a pliers and pull the plug out of the block. Remove all oil gallery plugs.

Engine parts can be cleaned several different ways. Iron or steel parts can be soaked in a tank full of a heated alkaline solution (i.e., hot-tanked). This will remove oil, sludge, hard, baked-on carbon deposits, and mineral deposits in the coolant passages. Never put any aluminum parts in a hot tank—the caustic solution will corrode aluminum.

Many shops have cold solvent parts washers. Small to medium size parts can be placed in the washer and sprayed with solvent. Brushes or scrapers can be used to remove stubborn deposits from the parts.

Engine parts can also be cleaned in a thermal cleaner. These are actually large ovens that heat parts to temperatures between 650 and 800° F (343 and 427° C) to oxidize the contaminants. After the thermal cleaning process, the ash is removed by shot blasting or washing the parts. The temperature inside a thermal cleaner can also be reduced to clean aluminum parts without damaging them.

Regardless of which cleaning method is used, always perform a careful inspection of oil passages in the cylinder block, head, crankshaft, and all other parts. This is especially important if the engine suffered a major failure like a spun bearing or a severely worn camshaft and lifters. Metal particles will become lodged in the oil passages and can be difficult to remove. Rod out all small diameter passages to make sure that they are unobstructed. Use a long, slender rifle brush to thoroughly clean oil galleries.

Task C.2   Visually inspect engine block for cracks, corrosion, passage condition, expansion and gallery plug hole condition, surface warpage, and surface finish and condition; determine necessary action.

After disassembly and thorough cleaning you must make a careful visual inspection of the block for cracks. Any areas that are in question can be further inspected by a machine shop using dyes and magna-fluxing processes. Special attention to areas that may have corrosion that would affect good sealing on reassembly is important. Areas that seal coolant such as water pump passages and expansion plug holes are vulnerable. It would be important to look through the expansion plug holes to the cylinder walls on cast iron blocks. If a lot of pitting is present it may be necessary to replace the block to avoid internal leaks after over-boring. Be cautious of conditions like this when the cooling system condition is very poor and rusty. Next, inspect oil galleries for sludge or casting flash that may break loose and damage the engine on reassembly. Check all gasket surfaces for pitting or warpage and make sure that they meet surface conditions set by the manufacturer. If in doubt enlist the advice of a good machinist. Most head gaskets used in late model engines are not installed with any kind of sealer so the surface condition and cleanliness are critical to successful sealing. When checking the head surfaces or deck of the block for straightness, if you can measure any warpage with a straight edge and feeler gauge some action will almost certainly have to be taken.

Task C.3   Inspect and repair damaged threads where allowed; install expansion and gallery plugs.

Inspect all threaded bolt holes in the cylinder block, especially the cylinder head bolt holes. Check holes for sludge or debris that may have been missed during cylinder block cleaning. Run the appropriate size tap into each of the cylinder head bolt threaded holes to make sure that the threads are clean. Then put a few drops of oil on a head bolt and thread it into each hole by hand.

If the threads in a bolt hole are damaged, the hole may be drilled to a larger size and rethreaded. Then a thread repair insert or heli-coil can be installed in the oversize hole. The result is a threaded hole that is the same size as the original. Different types and brands of thread inserts are available. In most cases, however, the insert is threaded onto a special installation tool, coated with thread locking compound, and then threaded into the oversize hole. The installation tool is then removed and a hammer and punch are used to break off a tang at the bottom of the insert.

Inspect oil gallery plug threaded holes for dirt or damage. Do not run a tap very far into these holes since they usually have tapered pipe threads. Run a long brush (called a rifle brush) down the oil galleries to make sure that all debris has been removed. Then coat the new oil gallery plugs with teflon tape or an oil-resistant sealer and thread them into the block. Do not overtighten the plugs. If there are small expansion plugs at the ends of the oil galleries, coat the edges of new plugs with an oil-resistant sealer and then drive the plugs into the block. Use a cold chisel and hammer to cross stake the end of the bores after the expansion plugs are installed.

Clean out expansion plug bores with emery cloth before installing new plugs. If a bore is damaged, it may be repaired by boring it to the next specified oversized plug. Oversized expansion plugs are stamped with the letters OS. Before installing a new plug, coat the sealing edge with a nonhardening, water-resistant sealer. Drive the plugs into the block using the proper special driving tool. Make sure that the plug goes into the bore squarely to prevent leaks.

Task C.4   **Inspect and measure cylinder walls; remove cylinder wall ridges; hone and clean cylinder walls; determine need for further action.**

If the ring ridge at the top of each cylinder has not already been removed, remove it now using a ridge reamer. While using the ridge reamer, be careful to avoid marking the cylinder walls below the ring ridge. Do not remove any metal from the cylinder wall below the ring ridge. Failure to remove the ring ridge may cause piston ring lands and/or the top compression ring to crack or break after the engine is assembled and started.

Use a dial bore gauge to measure the cylinder diameter in three vertical locations. These locations are just below the ring ridge at the top of the cylinder, in the center of the ring travel, and just above the lowest part of the ring travel. Cylinder taper is the difference in the cylinder diameter at the top of the ring travel compared to the diameter at the bottom of the ring travel.

In each of the three vertical cylinder measurement locations, measure the cylinder diameter in the thrust direction and in the axial direction. Cylinder out-of-round is the difference between the cylinder diameter in the thrust and axial directions. If cylinder out-of-round exceeds specifications, rebore the cylinder.

If cylinder wear, out-of-round, and taper are within specifications, the cylinders may be deglazed. Very mildly worn cylinders should be deglazed with a brush hone, which removes material very slowly. Moderately worn cylinders (still within specifications) may be deglazed with 220 or 280 grit stones installed on a cylinder hone. When the honing operation is completed, the cylinders should have a 50° to 60° crosshatch pattern. After deglazing, the cylinder should be cleaned with hot, soapy water and a stiff-bristle brush. Ordinary solvent will *not* remove grit from pores in the cylinder wall—use hot soapy water. The bores are clean when a clean, lint-free cloth is used to wipe them and the cloth does not get dirty. When the bores are clean, rinse the block and dry it thoroughly. Coat all machined surfaces with a light coating of the manufacturer's recommended engine oil.

If one cylinder requires reboring, most manufacturers recommend reboring all the cylinders to the same size. Cylinder reboring usually is done with a specialized piece of equipment called a boring bar. Cylinders must be honed after boring. A honing machine is usually used for this. After cylinder honing, the same procedure for block cleaning should be followed as previously discussed in cylinder deglazing.

Task C.5   **Inspect crankshaft for end play, straightness, journal damage, keyway damage, thrust flange and sealing surface condition, and visual surface cracks; check oil passage condition; measure journal wear; check crankshaft sensor reluctor ring.**

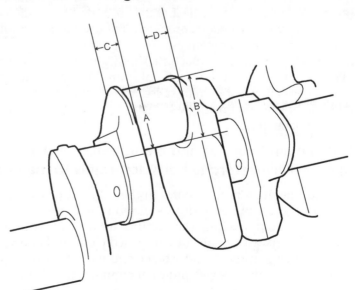

Crankshafts must be inspected at several different points to ensure dependable engine performance.

End play is the amount the crank moves fore and aft in the block. One of the crank shaft bearings provides a thrust surface that controls fore and aft movement. Crank wear or bearing wear can allow excessive end play which can affect clutch operation in manual transmission equipped vehicles, belt alignment problems, starter engagement problems and even automatic transmission pump or converter damage. This measurement is made when the crank and main bearings are assembled in the block by using a dial indicator in a horizontal or parallel mount to the crank and measuring total fore and aft movement.

A crank can be measured for straightness by placing it in a set of V-blocks and using a dial indicator to measure the runout of the main bearing surfaces. Another good hint of a bent crank is excessive main bearing wear on one or two journals compared to others. This is not an indicator by itself as each journal's bearing clearances can be a factor as well.

The crankshaft bearing surfaces must be inspected for cracks and wear grooves. The crank must be replaced, resurfaced and or repaired if problems are found here. Clean out the oil passages with thin rods or small gallery brushes to be sure no restrictions are present.

Using the provided diagram we need to make some measurements to determine if it is necessary to resurface the crank. Using a micrometer we need to determine the difference between point A and B. This will tell us if the bearing surface is tapered. Comparing A to C and B to D will tell us if we have an out of round condition. If the crank is even you can determine bearing clearance with a standard bearing by comparing your measurements to the specs for the crank. If the crank is not larger than the minimum spec it may be necessary to resurface the crank to the next undersize and use appropriate bearings.

Task C.6   Inspect and measure main bearing bores and cap alignment and fit.

Check main bearing bore alignment before bearing caps are installed. The block should be resting on a flat surface, *not* hanging from an engine stand by the flywheel end. Lay a straightedge across the bores and check alignment using feeler gauges. If bearing bores are not aligned, the block can be line bored.

Check the bearing cap and cylinder block mating surfaces for nicks and burrs. These can be removed using a file.

When measuring main bearing bore diameters, the bearing caps must be installed and the bolts properly torqued. Check bore diameter in three directions. The vertical measurement should not be larger than any of the others. A larger vertical reading indicates the bore is stretched. Out-of-round measurements less than 0.001 inch (0.025 mm) are acceptable, provided that the vertical reading is not the largest.

Improper bore alignment and bore dimensions can be corrected by line boring. This operation, performed by machine shops, involves removing the bearing caps and planing a small amount of material from the cap surface that mates with the cylinder block. The caps are then reinstalled and torqued to specifications. A specialized piece of equipment called a line hone is used to "true up" the main bearing bores to their original diameters.

Task C.7   Install main bearings and crankshaft; check bearing clearances and end play; replace/retorque bolts according to manufacturer's procedures.

Clean the main bearing bores in the cylinder block and bearing caps with solvent and allow the bore surfaces to dry. Do not oil the bores. Handle new bearing inserts carefully—avoid touching the bearing surface with your fingers. Wipe the back of the bearing inserts with a solvent-dampened cloth and allow the inserts to dry. Install the bearing inserts in the cylinder block and main bearing cap bores. The upper bearing halves are usually grooved and each contains an oil supply hole. Make sure that the oil hole in the bearing aligns with the oil hole in the bearing bore. Make sure that the tab on each bearing insert fits tightly in its bearing bore notch.

Carefully lay the crankshaft in the cylinder block.

To measure bearing clearance, install a strip of Plastigage across each journal. Then install the bearing caps and bolts, tightening the bearing cap bolts to the specified torque. Remove the bearing cap bolts and the bearing caps. Compare the width of the crushed Plastigage strip on the bearing journal to the scale provided on the Plastigage package to determine the bearing clearance.

Crankshaft end play may be measured by inserting a feeler gauge between the crankshaft thrust journal and the thrust lip on one of the main bearings. On some engines, a dial indicator is used to measure crankshaft end play while moving the crankshaft with a pry bar. Excessive end play may cause premature bearing wear or noise as the crank chucks back and forth in the block.

If bearing clearance and end play are within specifications, remove the main bearing caps and the crankshaft. Install a rope-type or split lip seal in its grooves in the block and rear main bearing cap. Oil the main bearing inserts, lay the crankshaft in the block, and install the main bearing caps.

Install the bearing cap bolts and torque them to specifications. Some engines are fitted with torque to yield (TTY) main bearing cap bolts that must be replaced after being loosened. Check the manufacturer's service manual to determine whether the engine you are servicing has TTY main bearing cap bolts.

Task C.8 Inspect camshaft bearings for unusual wear; remove and replace camshaft bearings; install camshaft, timing chain, and gears; check end play.

Inspect the camshaft bearings for scoring, roughness, and wear. Camshaft bearings or bearing bores should be measured at two different locations with a telescoping gauge. Measure the camshaft journals with a micrometer, and subtract the journal diameter from the bearing diameter to obtain the clearance. If the wear exceeds specifications, replace the bearings.

The type of tool needed to remove and install the camshaft bearings depends upon the engine design. Most overhead valve (OHV), camshaft-in-block engines will use a camshaft bushing driver and hammer. The right size mandrel is selected to fit the bearing. Turning the handle tightens the mandrel against the camshaft bearing. Then the bearing is driven out by hammer blows. The same tool is used to replace the bearings. Some overhead camshaft (OHC) engines require a special puller/installer.

Never attempt to remove and install camshaft bearings in an OHC cylinder head using a bushing driver. The bearing supports may be bent or broken due to the hammer blows, especially on aluminum heads.

When installing cam bearings, it is very important that the bearing insert be properly positioned in the bearing bore. Be absolutely sure that any oil hole(s) in the bearing insert align with oil supply passages in the bearing bore. This may mean that the insert is positioned toward the front of the bore, the back of the bore, or even the center of the bore. The position is not important so long as the oil holes line up.

Many overhead cam engines do not have removable camshaft bearings—the camshaft journals run directly against bearing bores machined into the aluminum cylinder head. Inspect the bearing surfaces on these engines for scoring, roughness, and wear. Measure the bearing bores at two different locations with a telescoping gauge. Measure the camshaft journals with a micrometer, and subtract the journal diameter from the bearing diameter to obtain the clearance. If the wear exceeds specifications, replace the cylinder head.

When the camshaft bearing bores are machined into the cylinder head, the bearing caps should be removed and a straightedge positioned across all the bearing bores. Insert a feeler gauge between the straightedge and each bearing bore to measure any misalignment. Misalignment indicates that the cylinder head may be warped. Have a machine shop check for this. Cylinder heads can sometimes be straightened. Severely warped heads should be replaced.

Task C.9 **Inspect auxiliary (balance, intermediate, idler, counterbalance, or silencer) shaft(s), drives, and support bearings for damage and wear; determine necessary action.**

Balance shafts are found on many 4- and 6-cylinder engines. Some rotate at crankshaft speed; some rotate at twice crankshaft speed. Removal and installation procedures vary widely; refer to the manufacturer's service manual for service information.

Some balance shafts are mounted to the bottom of the cylinder block and are chain driven off the crankshaft. Some are gear driven by a large gear machined into a disc that is part of the crankshaft. Belt-driven balance shafts are often mounted inside the bottom of the cylinder block, much like camshafts on camshaft-in-block engines. On some popular V-type engines, a balance shaft is mounted in the cylinder block directly above the camshaft and gear driven off the camshaft.

Once removed, balance shafts should be checked for runout with the same procedure used for measuring camshaft runout. The balance shaft journals should be measured for taper with the same procedure for measuring crankshaft journal taper. When the balance shafts are installed, they must be properly timed to the crankshaft or severe engine vibration may occur upon engine startup.

Task C.10 **Inspect, measure, service, repair, or replace pistons, piston pins, and pin bushings; identify piston and bearing wear patterns that indicate connecting rod alignment problems; determine necessary action.**

Inspect pistons for cracks and damage from overheating. Pistons with these conditions should be replaced. Inspect the piston skirts for uneven wear. Wear on the edges of the piston skirt next to the wrist pin hole may be caused by a bent or twisted connecting rod.

Clean the piston ring grooves using a ring groove cleaning tool. Be careful to avoid removing material from the bottom of the groove. Make sure that oil drain holes at the bottom of the oil ring groove are not obstructed.

To measure piston ring side clearance, insert a new ring in the piston groove and position a feeler gauge between the ring and the groove. If ring side clearance is excessive, the piston should be replaced.

If the piston passes the inspections mentioned previously, check for a worn wrist pin bushing. Clamp the connecting rod body lightly in a vise (use soft jaw covers) and attempt to rock the piston against the connecting rod sideways (at a right angle to normal piston/connecting rod motion). If any play is noticeable, the piston wrist pin bore, wrist pin, or connecting rod small end bushing are worn and the piston and connecting rod must be separated. If no play is noticeable and the piston is to be reused, the components need not be separated.

Pistons must be fitted to their cylinder bores. If piston-to-cylinder wall clearance is excessive, piston slap may be noticeable. If there is not enough clearance, piston scuffing will occur. Check piston clearance by measuring the cylinder bores and the piston diameters.

To measure piston diameter, position a micrometer to contact the piston thrust surfaces (the surfaces at right angles to the piston, or wrist, pin bore). The exact measuring location varies according to manufacturer. Most manufacturers, however, specify a point about ¾ inch below the wrist pin centerline.

Task C.11 Inspect connecting rods for damage, alignment, bore condition, and pin fit; determine necessary action.

Inspect the connecting rods for cracks and obvious damage. Remove the cap nuts, caps, and bearing inserts. Check that the cap bolts are not loose in the rod body.

Inspect the bearing inserts for uneven wear. If the front and rear edges of a bearing are worn more than the center area, the rod may be bent or twisted. If the bearing inserts are worn more at the parting line areas, the rod big end bore may be stretched. Machine shops have special jigs used to check for bent/twisted connecting rods. If rod bend or twist exceeds specifications, replace the connecting rod.

Measure the connecting rod big end bore for taper, out-of-round, and proper bore size. If any of these dimensions are not within specifications, the connecting rod should be rebuilt or replaced.

On piston/connecting rod assemblies with free-floating wrist pins, remove the pin retainer circlips or snaprings and slide the wrist pin out of its bores. Take note of piston to rod orientation and separate the piston and rod. On piston/connecting assemblies with press-fit wrist pins, use the appropriate press and adapters to remove the wrist pin. Measure the connecting rod small end bore (or "eye") diameter. If bore diameter exceeds specifications, it may be possible to ream out the bore and install an oversize wrist pin. Some rod eyes have a pressed in bushing. If the bore is worn on this type of rod a new bushing can be installed. The new bushing must be reamed to fit the wrist pin.

If the rod beam has an oil squirt hole, make sure that the passage from the hole to the big end bore is not obstructed.

Task C.12 Inspect, measure, and install or replace piston rings; assemble piston and connecting rod; install piston/rod assembly; check bearing clearance and side play; replace/retorque fasteners according to manufacturers' procedures.

When new piston rings are installed, ring end gap must be checked and adjusted. Compress a ring just enough so it fits in a cylinder. Insert a piston in the cylinder upside down (crown end first) and push the ring down near the bottom of the cylinder. The ring should be about 0.5 inch from the bottom. Use a feeler gauge to measure the gap where the piston ring ends meet. If the gap is too small, the ring ends must be filed to increase the gap. If the gap is too large, the wrong rings were selected or the cylinder was bored/honed incorrectly.

If the pistons and connecting rods were separated, reassemble them. Position the rod eye inside the piston, making sure that the parts are oriented correctly. Most pistons have a notch or arrow in their crown that must point toward the front of the engine. Connecting rod orientation varies according to the engine manufacturer. Refer to notes made during disassembly or check the service manual for information. On engines with free-floating wrist pins, dip the pin in engine oil and install it in its bore. Install new circlips or snaprings. When installing press-fit wrist pins, many manufacturers recommend that the rod eye be heated before pin installation. Some manufacturers recommend that the piston be heated in a piston heater, as well. When components are heated as necessary, use a press and the appropriate adapters to install the pin.

When installing the piston rings on the piston, install the oil rings first. Follow instructions that come with the rings to position the upper and lower rail gaps with respect to the expander gap. Oil ring rails can be spiraled into their slots. When installing compression rings, make sure that any marks stamped into the ring are facing upward. Install the bottom compression ring first and then the top compression ring using a piston ring expander. Do *not* spiral compression rings onto the piston. This can bend the rings, causing them to resemble a lock washer.

Clean the bearing bores in the connecting rods and rod caps and allow the bore surfaces to dry. Do not oil the bores. Handle new bearing inserts carefully—avoid touching the bearing surface with your fingers. Wipe the back of the bearing inserts with a solvent-dampened cloth and allow the inserts to dry. Install the bearing inserts in the rod and cap bores. If the rod has an oil squirt hole, make sure that the oil hole in the bearing aligns with the oil hole in the rod body. Make sure that the tab on each bearing insert fits tightly in its bearing bore notch.

Before installing the piston/connecting rod assembly in the block, install short lengths of fuel line or "chopsticks" over the rod bolts and position the piston ring end gaps according to the manufacturer's instructions. If no instructions are given, it is common practice to position the gaps as follows:

- Oil ring expander gap facing the front of the engine, directly above the wrist pin centerline.
- Upper oil ring rail gap 45° to one side of the expander gap.
- Lower oil ring rail gap 45° to the other side of the expander gap.
- Bottom compression ring gap on the left side of the piston (90° from the wrist pin).
- Top compression ring gap on the right side of the piston (90° from the wrist pin).

Rotate the crankshaft to position the crank pin for the piston/rod being installed at bottom dead center (BDC), install a ring compressor on the piston, and slide the piston/rod into the block. Push the rod body against the crank pin, remove the rubber protective hose, and temporarily install the rod cap and nuts. Install the remaining piston/rod assemblies.

To measure bearing clearance, remove the rod cap and install a strip of Plastigage across the crankshaft journal. Then install the bearing cap and nuts, tightening the nuts to the specified torque. Remove the bearing cap nuts and the bearing caps. Compare the width of the crushed Plastigage strip on the bearing journal to the scale provided on the Plastigage package to determine the bearing clearance.

Measure the side clearance at each connecting rod by inserting a feeler gauge between the side of the connecting rod and the edge of the crankshaft journal. If side clearance exceeds specifications, the sides of the connecting rod or crankshaft journal are worn.

Task C.13 Inspect, reinstall, or replace crankshaft vibration damper (harmonic balancer).

Special tools are required to remove and install the vibration damper. Using a regular gear puller to remove the vibration damper will damage the damper.

Inspect the rubber between the inner hub and outer inertia ring on the vibration damper. If this rubber is cracked, oil-soaked, deteriorated, or protruding from the damper, replace the damper. If the inertia ring on the damper is loose or has shifted forward or rearward on the hub, replace the damper. Inspect the vibration damper hub for cracks or a damaged keyway. If either of these conditions is present, replace the damper. Inspect the seal contact area on the hub for a wear groove or scoring. If either of these conditions is present, replace the damper or install a sleeve on the hub to provide a new seal contact area.

Task C.14 Inspect crankshaft flange and flywheel mating surfaces; inspect and replace crankshaft pilot bearing/bushing (if applicable); inspect flywheel/flexplate for cracks and wear (includes flywheel ring gear); measure flywheel runout; determine necessary action.

Inspect the crankshaft flange and the flywheel-to-crankshaft mating surface for metal burrs. Remove any metal burrs with fine emery paper. Be sure the threads in the crankshaft flange are in satisfactory condition. Replace the flywheel bolts and retainer (if fitted) if any damage is visible on these components. Install the flywheel, retainer, and bolts, and tighten the bolts following the torque and sequence provided by the engine manufacturer.

Inspect the flywheel for scoring and cracks in the clutch contact area. Minor score marks and ridges may be removed by resurfacing the flywheel. If deep cracks or grooves are present, the flywheel should be replaced.

Mount a dial indicator on the engine block or flywheel housing and position the dial indicator stem against the clutch contact area on the flywheel. Rotate the flywheel to measure the flywheel runout. If runout exceeds specifications, replace the flywheel.

Insert a finger in the inner pilot bearing race and rotate the race. If the bearing feels rough or loose, replace the bearing. Check a pilot bushing to verify that it is not loose. A transmission input shaft may be positioned in the pilot bushing to check for excessive play. If too much play exists, replace the bushing. A special puller may be used to remove the pilot bearing or bushing. The proper driver must be used to install the pilot bearing or bushing. Always verify that the transmission input shaft fits in a new bushing before attempting to install the engine (or transmission).

Inspect the starter ring gear for excessive wear or damage. On manual transmission flywheels, the starter ring gear is often replaceable. Remove the old gear by drilling a hole through the gear at the "root" between two teeth. Then position a cold chisel between the two teeth and strike it with a hammer. Take note of whether the gear has a chamfer on one side before removing it. To install the new gear, first heat it to about 400° F in an oven. Then slip it over the flywheel body and allow it to cool.

Task C.15 Inspect and replace pans and covers.

Always inspect the gasket mounting surfaces on sheet metal pans for warpage, and look for dished retaining bolt holes. Dished mounting holes and warped mounting surfaces must be straightened by hammering them flat again. Use a straightedge to verify that gasket surfaces are flat.

Gaskets are used to seal minor variations between two flat surfaces. Oil pan gaskets or rocker arm cover gaskets are usually manufactured from cork, rubber, or a combination of rubber and cork or rubber and silicone. Timing cover, water pump, and thermostat housing gaskets on older engines were made of specially treated paper. On newer engines, these same components are sealed to the block using synthetic rubber O-rings or silicone "spaghetti" seals. Paper gaskets should be coated with a nonhardening sealer prior to installation. Synthetic rubber and silicone seals are often installed without coatings or cements.

Task C.16 Assemble the engine using gaskets, seals, and formed-in-place (tube-applied) sealants, thread sealers, etc. according to manufacturers specifications.

Gasket sealing technology has changed dramatically in a relatively short period of time. This requires the technician to have a large base of knowledge on how to seal the many types of gaskets. The newest types of gaskets are the captive silicone or rubber gaskets that fit into a channel cut or cast into one of the components mating surfaces. These gaskets are very dependable and depend on surface tension and crush to seal. Many of these gaskets are reusable in low-pressure environments. Reusable gaskets for areas such as oil pans are a hybrid version of this gasket where the gasket is captive in a high density plastic and torque to yield bolts are used to tension the component. These gaskets are all installed dry on very clean surfaces. Cork or paper gaskets usually benefit from a light coat of sealer usually RTV (room-temperature vulcanizing) type. This type of sealer is used in some applications to actually replace the gasket altogether. In these situations the application is critical. Any air bubbles or gaps in the sealer will develop into leaks later on. This type of sealer comes in many varieties. In vehicles with computer controls the use of a sealer complimentary to oxygen sensors is important. The gases released during the vulcanizing process and for some time after will cause the oxygen sensor to fail. Another gasket-less environment is the anaerobic sealer that dries when oxygen is not present. This is used as a sealer in applications where two machined mating surfaces must seal and maintain a precision location or clearance when assembled.

Seals are usually in the form of lip seals that fit over shafts allowing the sealed component to move. These seals depend on the oil they seal to provide lubrication. Due to this fact these seals have the highest failure rate. A rear main seal is an example of this type of seal whether it be full circle or split type. The surfaces these seals mate with are subject to wear that leaves a groove over time. These grooves can cause repeated replacements if they are not removed, the component replaced, or a repair sleeve installed.

In all cases where sealing is the goal, cleanliness is critical to success. It is important to use cleaners that do not leave residue behind. Chlorinated cleaners or denatured alcohol are the best choices.

In many applications a thread sealer is called for to control coolant or oil where bolts pass into areas containing fluids. It is important to use the sealer designed for the application as some sealers may be oil or coolant soluble.

D. Lubrication and Cooling Systems Diagnosis and Repair (8 Questions)

Task D.1   **Diagnose engine lubrication system problems; perform oil pressure tests; determine necessary action.**

In most cases diagnosis of engine lubrication problems is more forensic than preventive. Oil is critical to lubrication as well as internal engine cooling. When a component does not receive adequate oil it will fail quickly. The most common causes of oil related failures is plain old dirty oil. If the chain has worn a groove in the guide, it can begin to adhere to internal engine surfaces creating sludge or a varnish that can restrict oil passages or drains. Lifters and valve lash adjusters are particularly sensitive to these problems due to very small passages inside them. Something as simple as a stuck open thermostat can cause the engine to not reach operating temperature promoting oil sludging. This sludge can find it's way back into the oil pan and restrict the oil pump pick-up tube.

On the diagnostic side of the oil system, technicians are often called on to determine the cause of low oil pressure readings. This calls for removing the oil pressure sender or switch and checking oil pressure with a mechanical gauge. Manufacturers have their own specs and different rpm readings that they want tests conducted at, so don't assume that 20 psi is low unless the manufacturer tells you so. In the case of complete loss of oil pressure there are a couple of common causes. The pressure relief valve in the oil pump can stick open causing by-pass within the pump. The other common cause is blockage of the oil pump pick-up with sludge. Both are the result of poor maintenance. Low oil pressure can be caused by the aforementioned causes, excessive bearing clearances (high mileage engines fall victim to this), an oil filter that was installed too tight, or some other internal leakage. High oil pressure is usually only caused by a stuck closed pressure relief valve, higher viscosity oil than called for, or installing the incorrect oil pump. This will usually only occur in a situation where a high volume or high pressure pump was substituted for the original design.

If all pressures meet specs and the gauge was reading out of spec, you must consider and test the electrical components or the oil pressure warning system.

Task D.2   **Disassemble, inspect and measure oil pump (includes gears, rotors, housing, and pick-up assembly), pressure relief devices, and pump drive; determine necessary action; replace oil filter.**

Inspect the oil pump pressure relief valve for sticking and wear. If this valve sticks in the closed position, oil pressure will be too high. A pressure relief valve stuck in the open position results in low oil pressure.

On rotor type oil pumps, measure the thickness of the inner and outer rotors with a micrometer. When this thickness is less than specified on either rotor, replace the rotors or the oil pump. The following oil pump measurements should be performed with a feeler gauge:

- Measure pump cover flatness with a feeler gauge positioned between a straightedge and the cover.
- Measure the clearance between the outer rotor and the housing.
- Measure the clearance between the inner and outer rotors with the rotors installed.
- Measure the clearance between the top of the rotors and a straightedge positioned across the top of the oil pump.

On gear type oil pumps, inspect the gears and housing for scoring and excessive wear. When reassembling a gear type pump be sure to align any match marks stamped onto the pump gears.

Oil pump pick-up tubes should be carefully cleaned or preferably replaced if any sludge is present. While it would seem that oil filter replacement is a no-brainer, don't forget to rule out that a filter installed too tight can cause oil delivery problems.

Task D.3   Perform cooling system tests; determine necessary action.

Cooling systems have to perform several functions and so they require tests for all of those functions. Probably the first and most important thing it must do is actually hold coolant. Leaks are the most common complaint with cooling systems. Since the system experiences considerable expansion and contraction, components will eventually become stressed and develop cracks or leaks. The first level or leak diagnosis is to visually inspect the system to find leaks. Many undetected leaks are "cold" leaks, which means that they only occur when the engine is cold and maybe only when it is running and cold. These are most often found at hose fittings where clamps that perform well when warm are too loose during cold operation. When visually inspecting for leaks, don't forget to look inside the vehicle for leaks from the heater core or the hoses that attach to it. A complaint of a sweet smell in the cabin or steam on the windshield is your tip off that the heater core or heater hoses are leaking into the HVAC system.

If after visual inspection the leak is not found, applying pressure to the system with a pressure tester is the next order of business. When applying pressure, do not exceed the systems maximum pressure by more than 2 pounds or you may create a few new leaks of your own. Don't forget to test the radiator cap. When a system does not hold pressure, it is not uncommon for coolant levels to look low because the coolant gets pushed out or into the recovery bottle and not drawn back into the system because a vacuum does not occur as it cools to draw the coolant back into the radiator. This can also be your leak as the coolant expansion may be more than the recovery tank can hold. Keep in mind too that any leak that keeps the system from holding pressure can cause this.

Internal engine leakage is probably the worst case scenario in coolant leaks. Many times a persistent coolant loss is caused by a leaking head gasket or crack in an engine casting. It is important for us to diagnosis this type of problem as early as possible because of other potential collateral damage that can occur. Many leaks are between the combustion chamber and the cooling system. These, if undetected, can result in piston damage (since coolant doesn't compress). Radiator or heater core damage is common. Most cooling systems run 13-18 psi, cranking pressure will easily be over 100 psi. It doesn't take much to blow the tanks off of late model plastic radiators. Internal leaks that aren't exposed to combustion pressures can fill up the crankcase with coolant and cause bearing damage. Internal leaks can be spotted by removing the dipstick or oil fill cap and looking for something like whitish brown whipped cream. These leaks are usually head gaskets or intake manifold gaskets on wet valley V-type engines. In situations where a combustion leak is suspected but not readily apparent, you can often find them with a combustion leakage test that uses a colored liquid that changes color when exposed to exhaust gases in the cooling system.

The cooling system must help to maintain consistent temperature. An engine is a heat pump but because it does not have very high thermal efficiency it makes more heat than can be turned into power. That heat is carried away to be exchanged with ambient air by the cooling system. The water pump moves the coolant through the engine and maintains adequate pressure in the block to keep bubbles from forming on castings that could create hot spots. The thermostat is a controlled restriction that helps the engine warm up by virtually stopping coolant flow to the radiator until a prescribed temperature opens it. The thermostat may open and close during engine operation depending on ambient temperature and engine load. The radiator and heater core are air to water heat exchangers. The heater core is used to provide warm air inside the cabin and the radiator is used to remove extra heat from the engine. Late model cars have very carefully chosen radiators to help them achieve operating temperature quickly. It does not take very much restriction of either airflow or coolant flow to cause a drop in radiator efficiency that results in overheating when under loads. The radiator's efficiency can be tested by measuring the difference between inlet and outlet temperatures. A good rule of thumb is that a 40 degree drop is normal; the radiator may be restricted if that number is significantly higher or lower. Restriction can be in the form or debris collected in the radiator fins or deposits inside the tubes slowing or restricting coolant flow.

Task D.4   Inspect, replace, and adjust drive belts, tensioners, and pulleys.

Inspect the accessory drive belts for condition and tension. On conventional V-belts, the sides of the belt are the friction surfaces, so check the sides for cracks, glazing or loose cord material. Replace a belt showing any of these conditions. If a V-belt is severely worn, it may contact the bottom of the pulleys. Replace severely worn belts. If a belt is severely worn on just one side, check pulley alignment. V-belt pulleys must be aligned within 1/16 inch (1.6 mm) per foot of belt span. If pulleys are not aligned, check for loose accessory mounting bolts, missing spacers, or bent brackets.

V-belt tension can be checked using a variety of special testers. With one type of tester, the tool is placed over the belt at the center of a belt span. Squeezing the tool handles causes the tool dial to display belt tension, usually in pounds. Belt tension can also be checked by measuring the amount of belt deflection with a ruler. Use your thumb to press on the belt at the middle of a span while holding the ruler next to the belt. If belt tension is correct, the belt should deflect 0.5 inch (12.7 mm) for every foot (30.5 cm) of belt span.

A moderately loose or worn belt may cause a squealing noise when the engine is accelerated. A severely worn or loose belt may cause a discharged battery, engine overheating, or a lack of power steering assist. An overtightened belt may fail suddenly, or damage the alternator front bearing. An overtightened belt can also cause the upper half of the crankshaft front main bearing to wear prematurely.

When repositioning an accessory device (alternator, power steering pump, etc.) to adjust belt tension, always look for the pry points provided by the manufacturer. Some devices have slots for inserting a large screwdriver or pry bar. Others have built-in square holes to accommodate a 0.5 inch breaker bar or ratchet. Never pry on a power steering pump housing to tighten the drive belt. These housings are not meant to withstand such abuse and will be damaged, possibly causing a fluid leak.

Inspect Serpentine belts for missing ribs, wear on outside edges, cracks closer than 2 inches apart and excessive glazing. Any of these require immediate replacement.

Serpentine or V-ribbed belts must also be properly tensioned, but these belts are usually fitted with automatic tensioners. The tensioner automatically adjusts belt tension as the V-ribbed belt stretches. These tensioners often have built-in wear indicator scales. So long as an arrow on the tensioner is located between two lines, the belt is not excessively stretched. When the arrow moves outside the lines, the belt must be replaced.

Many vehicles are equipped with plastic tensioner and idler pulleys. It is very important to inspect these for wear. They will allow belt tension to drop or even damage the belt when they wear.

Task D.5   Inspect and replace engine cooling and heater system hoses and fittings.

Check all cooling system hoses for loose clamps, leaks, and damage. Look for cracks, abrasions, bulges, and swelling. Check for hard spots due to heat damage from close proximity to exhaust system components. Also look for shiny spots caused by contact with accessory mounting brackets or other components. These spots may indicate weak spots that could cause a hose to burst. Check the hoses for soft or gummy areas due to contact with engine oil, power steering fluid, or automatic transmission fluid.

Squeeze each hose along its entire length to check for hard or soft areas. Also listen for crackling or crunching noises while squeezing which would indicate that the reinforcing fabric is faulty or the inner liner has deteriorated. Lower radiator hoses often contain a steel spring to prevent the hose from collapsing, so you may not be able to perform the squeeze test.

When in doubt about a hose's condition, remove it and inspect the inner liner. If the liner is cracked or otherwise deteriorated, replace the hose.

Be careful when removing a faulty hose. Aggressive twisting and pulling can damage a heater core or radiator tank. If the hose is stuck to the fitting, slit the end of the hose to make removal easier.

When installing a new hose, make sure that it fits properly. Avoid twisting or stretching the hose. A hose that is too short may fail when the engine shifts during acceleration.

Task D.6   Inspect, test, and replace thermostat, bypass, and housing.

The thermostat may be tested after it has been removed from the engine. Submerge the thermostat in a pan of water and put a thermometer in the water. Suspend the thermostat and the thermometer above the bottom of the pan. Allowing them to lay on the bottom of the pan will not allow an accurate test. Heat the water while observing the thermostat valve and the thermometer. The thermostat valve should begin to open when the temperature on the thermometer is equal to the rated temperature stamped on the thermostat. Replace the thermostat if it does not open at the rated temperature.

Always replace a thermostat with one having the correct temperature rating. Do not install a "hotter" thermostat in an attempt to speed up engine warm-up time. The engine will warm up at the same rate, but operate at a higher temperature. Do not remove a functioning thermostat from an engine that is overheating. While the engine may stop overheating, coolant will flow through the engine too quickly to absorb heat adequately. Hot spots will develop in the cooling system, especially in the cylinder heads. A cracked head can result.

Be sure to install a thermostat in the correct direction and orientation. Many thermostats have an arrow indicating which way coolant should flow through the thermostat. Some thermostats have a vent hole, "jiggle" pin, or check ball assembly mounted toward the edge of the mounting flange. This device, which allows trapped air to pass through the thermostat, must be oriented properly. In most cases the device must point upward. Check the engine manufacturer's service manual for instructions.

Inspect the thermostat housing and bypass hose (if equipped) for cracks, deterioration, and restrictions. Thermostat housings are often made of sheet metal or a light alloy that corrodes rapidly when coolant is not changed at the recommended intervals. Replace a deteriorated thermostat housing or bypass hose.

Task D.7   Inspect coolant; drain, flush, and refill cooling system with recommended coolant; bleed air as required.

Cooling system service is a topic approached differently by technicians in different repair venues. We will offer information that should be generic to all repair technicians. ASE workshop participants must all agree on the content of the test so manufacturer specific items will not appear in the test unless they are considered industry standard. It is very important to keep this in mind when taking the test.

When testing coolant, there are many methods to arrive at results but the ultimate results are the same. We want to know the protection levels of the coolant for freezing, boiling, PH, corrosion protection and in some vehicles, nitrites. Let's take a quick look at each area.

Freezing and boiling protection are linked for the most part. Most manufacturers agree that a mixture of 50% water to 50% coolant provides the best of both worlds in this area and the best component protection. All manufacturers will also agree that you should be sure to use the correct coolant in the vehicle without mixing coolant types or changing to one not designed for the vehicle.

PH is a measurement of the acidity or alkaline qualities of the coolant. As coolant becomes older, it drops toward the acid end of the PH scale. Most Asian vehicles aim for around 7–9 and most American and European vehicles aim for 8–9.5 on the scale. Low PH readings can be deteriorated antifreeze condition or a water heavy blend as water is more toward the acidic side than a coolant mix is. Very high numbers can be caused by over adding antifreeze or corrosion packages during service.

In normal use vehicles with high output ignition systems, particularly DIS systems, will cause the coolant to become electrically charged which promotes debris in the system to adhere to metal parts and can cause radiator restriction. This can really only be corrected by replacing the antifreeze or reversing the charge in the system by using some ionizing coolant recovery systems.

Corrosion protection is added when servicing the cooling system with recovery/recycling equipment and is in the antifreeze to begin with. This is a difficult area to test and a bone of contention with manufacturers who do not support coolant recycling.

The last area that will become more critical as more diesel vehicles enter the consumer market is nitrites. When out of balance they cause small bubbles to collect on castings while the engine is running. These bubbles act like little cutters over time and carve into the casting. The vibration inherent in the diesel combustion process has been known to cause bubbles in the system to create leaks in cylinder walls. There are test strips available that detect the level of nitrites and currently only a couple of manufacturers have any specification for them.

On some vehicles, air pockets tend to develop in the cooling system as it is filled. If these air pockets are not bled off, engine overheating and even a cracked cylinder head can result. Some engines have a bleed fitting installed on the thermostat housing or an engine coolant passage to release trapped air. Loosen this fitting until all air is removed. On engines that do not have a bleed fitting, locate the highest point in the cooling system. If this point is a hose connection, loosen the hose to bleed off air. If a high point to bleed the system is not available, there are tools available that help to push the coolant from the bottom up displacing any air locks before the vehicle is started.

Task D.8   Inspect and replace water pump.

Check the water pump for leaking hose connections, mounting gaskets, and seals. Slow or hard-to-find leaks may be easier to find if a cooling system pressure tester is connected to the radiator filler neck.

Locate and examine the vent, or weep hole in the water pump housing. The hole is usually in the underside of the housing, so use a small inspection mirror, if necessary. If the water pump seal is leaking, coolant will usually drip from the weep hole. A very slow leak may leave only coolant residue around the hole. Replace the pump if there is evidence of coolant at the weep hole.

A defective water pump bearing may cause a growling noise at idle speed. In some cases, the bearing starts to fail after being contaminated by coolant leaking past the pump seal.

With the engine shut off, grasp the fan blades or the water pump pulley, and try to move it from side to side. This will reveal any looseness in the water pump bearing. If there is any side-to-side movement in the bearing, the water pump should be replaced.

When replacing a pump, always compare the new pump to the old one. Two pumps may look very similar, but their impellers may rotate in opposite directions. In this case, the impeller blades will be shaped differently and installing the wrong pump will cause the engine to overheat.

On many engines, some of the water pump mounting bolts extend into the block water jacket. Be sure to use the specified sealant on these bolts or coolant may leak from the engine. The bolts may also seize in place, making future servicing difficult. Refer to the manufacturer's service manual for information to determine which bolts enter the water jacket.

Task D.9   Inspect, test, and replace radiator, heater core, pressure cap, and coolant recovery system.

Examine the radiator for obvious damage or defects. Look for bent fins and fins clogged with dirt, road debris, or insects. These conditions greatly reduce radiator efficiency and can cause engine overheating. If damage is not severe, bent fins can usually be straightened using a special comb made for this purpose. Dirt and insects can be removed using a stream of low pressure water or compressed air.

Check the radiator for leaks or damp spots. Cracked solder seams and corroded tubes in copper/brass radiators can allow coolant to leak very slowly without leaving puddles of coolant. The same is true for aluminum/plastic radiators with cracked plastic tanks and leaking tank gaskets. Hard to find leaks can be located by removing the radiator, plugging the inlet and outlet fittings, and pressurizing the radiator with a cooling system tester. Submerge the radiator in a tank of water and check for bubbles.

Check the radiator cap for corrosion and damaged or deteriorated gaskets. Check the radiator filler neck seat, too. If the cap gasket or filler neck seat is damaged, the cooling system may not pressurize enough to prevent boilover. Coolant will be forced out of the cooling system and onto the ground or into the coolant recovery tank, if the vehicle has one. The engine may overheat.

If the vehicle is equipped with a coolant recovery system, check the gasket at the very top of the radiator cap. If this gasket is missing or leaking, coolant may be forced into the recovery tank when the engine warms up, but may not be drawn back into the radiator during cool down. Check the tube that connects the radiator filler neck to the reservoir for kinks, damage, or loose connections. These conditions may also allow coolant to flow into the tank, but not return to the radiator. Check the coolant reservoir for cracks, loose fittings and other damage. Some recovery systems use a reservoir cap that allows a length of tubing to hang down into the coolant. If this hose is missing or damaged, coolant cannot return to the radiator when the engine cools down.

Task D.10   Clean, inspect, test, and replace fan (both electrical and mechanical), fan clutch, fan shroud, air dams, and cooling related temperature sensors.

On rear-wheel-drive vehicles, the engine cooling fan is usually mounted to the water pump shaft and belt driven off the crankshaft. Plain, direct-drive fan blade assemblies should be checked for loose mounting bolts, cracked blades, and loose rivets (if fan blades are riveted to a hub). A fan assembly that has any cracks should be replaced immediately. Fans with temperature sensitive clutches should be checked for bad bearings, leaking fluid, and seized or free-wheeling clutches. With the engine off, try to spin the fan by hand. It should spin smoothly with some resistance. If the bearing feels rough, or the fan spins without resistance, replace the clutch assembly. Grasp the fan blades and try to rock the fan from side to side. Too much play indicates a bad bearing and the clutch should be replaced. Check the bimetal coil on the front of the clutch. If it is wet or covered with dirt and grime, silicone fluid is leaking out of the fluid reservoir and the clutch should be replaced. To check for a seized fan clutch, start the engine and observe fan speed. When the engine is cold, the fan should not pull much air through the radiator, even when the engine is revved. As the engine warms up, fan speed (and noise) should increase noticeably. If fan noise and speed seem excessive, stop the engine and put paint marks on the fan pulley and the back of the fan clutch. Then hook up an engine timing light and start the engine. When the timing light is pointed at the back of the fan clutch, the paint marks should move relative to one another. If the paint marks stay together as engine speed is varied, the clutch is seized and must be replaced.

Front-wheel-drive vehicles are usually fitted with electrically powered fans. The fan operates only when necessary. Some fans have both high and low speeds; others have just one speed. Fan operation is usually triggered by coolant temperature and/or A/C system operation. In some systems, a temperature sensitive fan switch is threaded into a radiator tank or an engine part to sense coolant temperature. When coolant temperature approaches the upper limit of the normal operating range, the switch contacts close to turn on the fan. When coolant temperature drops to a preset value, the switch contacts open to turn off the fan. Some switches supply power or ground directly to the fan motor; others activate a relay which then powers the fan. In other systems, fan operation is controlled by the PCM, which obtains coolant temperature information from its engine coolant temperature sensor. When coolant temperature reaches a preset value, the computer activates the fan and continues to monitor coolant temperature. When the temperature drops to a preset value, the computer turns off the fan. Also note that some PCM controlled fans operate at variable speeds, dependent on load.

Coolant temperature switches can be normally open or normally closed, and sensor resistance specifications vary. Refer to the vehicle manufacturer's service manual to determine how the system you are working on operates.

When a vehicle is equipped with air conditioning, turning on the system usually activates the engine cooling fan automatically. In some cases, however, the fan is not turned on until refrigerant pressure reaches a preset value. Again, refer to the vehicle manufacturer's service manual for information.

Fan shrouds and air dams are an important part of a vehicle's cooling system. They should be in place and undamaged. The purpose of a fan shroud is to allow a round fan to create a low-pressure zone behind the entire radiator core. The purpose of an air dam is to create a high-pressure zone in front of the radiator. Both components encourage airflow through the radiator under different circumstances. The fan shroud supports good airflow at low speeds and the air dam supports good airflow at highway speeds. This would lead us to keep in mind that a vehicle that overheats at highway speeds may have a problem with the air dam and a vehicle that overheats in traffic could have a problem with the fan shroud. The most common problems are that they are broken or missing.

Task D.11   Inspect, test, and replace internal and external oil coolers.

Cooling systems may have additional coolers for engine oil, supercharger heat exchangers, transmission oil, or power steering oil. These fall into 2 categories internal and external. What we mean by internal are the coolers located inside the side tanks or the radiator. These are almost exclusively transmission coolers. They serve two purposes: to warm and maintain transmission fluid temperature. The extent of their failures is usually in the form of leakage either externally at cooler lines or internally which mixes transmission fluid with coolant.

The remainder of the category is external oil coolers. Engine oil coolers are the most common. These types of coolers function like radiators. They simply use the components on lubricating fluid to remove heat to an air to liquid heat exchanger. Most have a thermostatic control to keep them from passing oil into the cooler until it reaches a set temperature. That is because you can run fluids too cool and cause other problems due to an increase in viscosity or sludging. The other types of coolers are not relevant to this test but function in much the same manner. Keep in mind that after a component failure, coolers should be carefully flushed or replaced to avoid debris damaging the new component.

E. Fuel, Electrical, Ignition, and Exhaust Systems Inspection and Service (7 Questions)

Task E.1 **Inspect, clean or replace fuel and air induction system components, intake manifold, and gaskets.**

Fuel and Air induction system components as they relate to engine repair require us to look at maintenance and inspection of items when performing major engine repair. Let's start from the top down.

On many late model vehicles, the most significant external component of these systems is the airflow sensor. Although the air passing through the sensor is filtered, they still get dirty and can cause drivability problems for new or repaired engines that could be detrimental to ring seating or engine operating temperatures. Air meters often require cleaning with a suitable air intake solvent that removes debris and dirt that can shroud and insulate the temperature sensing bulb and cause the PCM to miscalculate air flow.

The next component in line is common to throttle body and port fuel injected vehicles and it is the throttle body. Typically we are looking for collections of dirt inside the throttle body, behind the throttle plates and in the idle air control passages. It is key that manufacturer's recommendations be followed when servicing these because the use of the wrong cleaner could destroy the whole part. It is generally safe to wipe them out with a shop towel and in most cases this will remove deposits. Carburetor cleaner and even some induction system cleaners could remove the coatings that are designed to resist dirt causing high or erratic idle speeds.

Next in the air tract is the intake manifold. When replacing or resealing intake manifolds, it is important to check them for warpage or corrosion around water passages. Dirt deposits can generally be removed in a parts washer prior to reassembly. Just be sure to remove any electrical switches before cleaning. Most intake manifold gaskets are intended to be installed without sealers and most are very much like cylinder head gaskets. Proper torque is critical to lasting gasket performance.

The fuel components that are on the engine have to be considered. The fuel rail and pressure regulator on fuel injected vehicles, the carburetor on older vehicles, the fuel injectors, and the related sensors and actuators all are included.

Carburetors, for the most part, saw their last application in 1995. Prior to 1985 most domestic and Asian vehicles were carbureted. Carburetors and the associated vacuum lines must be checked for leakage or deterioration.

The fuel rail and pressure regulator usually only require service when a failure occurs. During exchange or R&R operations it is wise to replace injector o-rings to avoid leaks that are time consuming after the job is completed. Pressure regulators usually have a vacuum source to the manifold either by a line or a passage in the throttle body on TBI engines. It is critical to make sure this line or passage is clear and in good condition to avoid over-fueling conditions caused by high fuel pressure.

Fuel injectors should be checked for evidence of leakage. O-rings should be replaced and attention should be given to any wiring that may have deteriorated.

The last area includes the sensors and actuators that relate to these systems. Many vehicles have air and coolant temperature sensors mounted on the intake manifold. Thermistor type air temp sensors should be inspected and replaced if deposits have built up on them. Cleaning often damages them. The throttle position sensor and manifold pressure sensors should be handled with care during service as should all electrical wiring. Idle speed control devices that can be cleaned should be serviced while the manifold is apart. Many idle speed control devices have air hoses that attach to them, These should be inspected for cracks or leaks that could allow unfiltered air to enter the air intake tract.

Task E.2   Inspect or replace air filters, filter housings, and intake ductwork.

Inspect the air filter housing and any ductwork directing fresh air to the housing. Ductwork often leads from the radiator support or inner fender to the air filter housing. Make sure that this ductwork is present and undamaged.

Check the air filter housing itself. Make sure that the housing is securely mounted and hose or duct connections are tight. On carbureted and throttle body injected (TBI) engines, make sure that the gasket between the air filter housing and the carburetor or TBI unit is present and in good condition. Check the housing lid or cover to make sure that it fits properly and any seals or gasket material are in good condition. Streaks of dust or other debris around a sealing area indicate that the seal is leaking. Make sure that all clips or wing nuts securing the lid or cover are present and working properly.

Check the air filter for damage or excessive dirt. Check the filter or filter housing for an instruction label. Follow the manufacturer's recommendations, if present, especially those for filter element replacement intervals.

Task E.3   Inspect turbocharger/supercharger; determine necessary action.

The turbocharger/supercharger and all its mounting brackets, heat shields, and ducting should be checked for damage. Replace or repair damaged or missing components.

Check the air intake side of the turbocharger/supercharger system for leaks. If there is a leak in the intake system before the compressor housing, dirt may enter the turbocharger and damage either the compressor or turbine wheel blades. When a leak is present in the intake system between the compressor wheel housing and the cylinders, turbocharger pressure is reduced.

Turbocharger/supercharger boost pressure may be tested with a pressure gauge connected to the intake manifold. Boost pressure should be tested during hard acceleration while driving the vehicle. Excessive boost pressure may be caused by a wastegate that is stuck closed, a leaking wastegate diaphragm, or a disconnected wastegate linkage. Reduced turbocharger boost pressure may be caused by a wastegate that is stuck open.

When diagnosing the cause of blue exhaust smoke on a turbocharged vehicle, first perform oil consumption diagnosis as though the engine was *not* turbocharged. While turbos are commonly blamed for excessive oil consumption problems, about half of the turbos returned under warranty are not defective. Refer to Task A.5 for more information about excessive oil consumption on turbocharged engines.

Task E.4   Test engine cranking system; determine needed repairs.

A battery's state of charge can be determined by measuring the specific gravity of the electrolyte solution, or "acid." To check specific gravity, remove the battery cell caps and use a hydrometer to withdraw some electrolyte from each cell. The electrolyte in a fully charged battery has a specific gravity of 1.260–1.270 when it is at a temperature of 80° F (27° C). If specific gravity is below 1.260, the battery should be charged before further testing. An important indication of battery condition is the spread or variation in specific gravity between cells. Specific gravity should not vary more than 0.050 from the lowest cell to the highest. If it does, the battery has internal damage and should be replaced.

On maintenance-free batteries, the cell caps are usually sealed to the battery case and cannot be removed. Some of these batteries have a built-in hydrometer, or "magic eye." Look down into the lens on top of the eye. If the eye appears to be green, the battery is sufficiently charged for testing. If the eye is dark, the battery should be charged before testing. If the eye is yellow or clear, the battery is low on electrolyte and should be replaced. Keep in mind that the built-in hydrometer measures specific gravity in only one cell, so a battery showing a green eye could still be defective.

Another way to determine a battery's state of charge is to measure open circuit voltage. First remove the battery surface charge by connecting a 50-ampere load across the battery terminals for 10 seconds. Then wait 10 minutes for the battery to stabilize. Disconnect both battery cables from the battery and use a voltmeter to measure voltage across the terminals. Open circuit voltage will be at least 12.6 volts on a fully charged battery *after* the surface charge has been removed. If open circuit voltage is below 12.4 volts, the battery must be charged before further testing.

The best way to determine a battery's ability to deliver power is to perform a load or capacity test. During a load test, a load tester is used to discharge the battery at one-half of its stated cold cranking ampere (CCA) rating for 15 seconds. Battery voltage is recorded at the end of this time, while the load is still applied. If the test is performed with the battery at about 70° F (21° C), a battery voltage reading of 9.6V or higher indicates that the battery is in good condition.

A discharged battery can be either fast charged or slow charged. Slow charging is always preferable if time is available to do this. Slow charging allows the chemical changes that take place during charging to occur throughout the entire thickness of the battery plates instead of on the surface of the plates only. Slow charging also lessens the chances that the battery will be overheated (and permanently damaged) during charging.

The correct slow charging rate for a battery depends on the battery's reserve capacity rating, but is usually 10 amperes or less. A battery with an 80 minute reserve capacity should be charged at 5 amperes for 10 hours or 10 amperes for 5 hours. If reserve capacity is about 150 minutes, the battery should be charged at 5 amperes for 20 hours or 10 amperes for 10 hours. Check specific gravity hourly while the battery is slow charging. When specific gravity does not increase from one check to the next, slow charging is complete.

Fast charging takes place at a much faster rate (often up to 30 amperes) and takes much less time (less than 2 hours) than slow charging. Electrolyte temperature and battery voltage should be monitored during fast charging. If electrolyte temperature reaches 125° F or voltage reaches 15V, reduce the charging rate. The correct fast charging rate depends upon battery capacity (ampere-hours) and specific gravity reading. A battery with a 55 ampere-hour capacity and a specific gravity reading of 1.200–1.225 (¾ charged) should be fast charged for about 35 minutes. If the same battery starts out at a specific gravity of 1.125–1.150 (dead), it should be fast charged for about 80 minutes. After verifying the battery is in good condition, inspect all of the battery cables. Cables that are frayed or corroded should be serviced or replaced as needed.

The next item we want to test is the starter. We are looking for current demand and for any unusual sounds. Depending on the engine and starter design most starters will require 150–200 amps to turn over the engine. If the starter is high there may be internal problems with it such as a dragging armature or worn bushings. Listen for noise or very loud cranking noises. Some vehicles require shims to properly place the starter. If the vehicle had shims in it be sure to replace them during reassembly.

Task E.5 Inspect and replace positive crankcase ventilation (PCV) system components.

The positive crankcase ventilation (PCV) system serves two purposes. First, it removes crankcase blowby and pressure. Second, it introduces the crankcase gases into the engine to be burned during the combustion process.

Inspect the system by looking for hoses that are cracked, swelled or kinked. The side of the system with the PCV valve in it is under vacuum. The side that connects to the air cleaner or air intake tube is the vent side. Air is drawn from the filtered intake side to ventilate the crankcase and then drawn through the PCV valve and into the intake stream.

Task E.6 Visually inspect and reinstall primary and secondary ignition system components; time distributor.

On engines fitted with a distributor, check for a cracked, worn out, or damaged distributor cap. Pull each spark plug wire from the cap (one at a time) and check for burned or corroded terminals. Check the spark plug cables for burned, pinched, cut, or oil-soaked insulation. Replace damaged cables. Remove the cap and check inside it. If the cap has excessively worn or corroded terminals, replace it. Check for carbon tracking and, if found, replace the cap. Check the high tension cable leading to the ignition coil. Ignition coils sometimes leak oil, which will soften and damage the cable. If oil is found, replace the ignition coil and the cable. Check the distributor rotor for a burned, pitted, or excessively worn contact.

On engines equipped with solid state ignition, check the centrifugal and vacuum advance mechanisms, if equipped. Inspect the reluctor or pole piece to make sure that it is not contracting the magnetic pickup or pickup coil. Replace damaged parts. The mechanical advance advances spark timing as engine speed increases. Check that the advance mechanism is not seized by grasping the rotor and attempting to turn it in the direction of rotor rotation. The rotor should move in the direction of rotation against spring pressure, but not the opposite direction. When the rotor moves, pivoted weights under the rotor or breaker plate should move outward. The vacuum advance unit controls spark advance in relation to engine load. To test the advance unit, connect a hand operated vacuum pump to the hose nipple on the unit and watch the breaker plate while operating the pump. The breaker plate should rotate in the direction opposite that of the distributor rotor as the pump is operated. If the unit will not hold vacuum, the diaphragm is damaged and the vacuum advance unit must be replaced. If the unit holds vacuum, but the breaker plate does not move, the plate may be seized. Check for a rusted pivot point or a foreign object (like a dropped point retaining screw) that may be jammed between the plate and the distributor housing.

Remove and inspect the spark plugs. Use the appropriate socket to prevent spark plug damage. Remove the plugs when the engine is cold, especially if the engine has aluminum cylinder heads. Removing plugs from an aluminum head when the engine is hot can damage the aluminum threads. If the plugs are in good condition, apply a small amount of antiseize compound to the spark plug threads. Then thread the plugs into the head by hand to avoid cross-threading. Torque the plugs to specifications.

On most, but not all, engines fitted with a distributor type ignition system, the distributor can be rotated to adjust spark timing. Vehicles produced since 1972 have an underhood emissions label that outlines the steps necessary to adjust spark timing. Follow these instructions. In a typical sequence, the engine is brought to normal operating temperature. Then the vacuum advance hose (if equipped) is disconnected and the hose is plugged. A stroboscopic timing light is connected to the #1 cylinder spark plug cable and the engine is started. Idle speed is adjusted to specifications, and then the timing light is pointed at a metal tab attached to the timing cover. If necessary, the distributor is rotated to cause a notch in the crankshaft pulley, harmonic balancer, or flywheel to align with a timing mark. On distributorless ignition systems inspect coil packs, plug wires and modules for signs of arcing; replace any damaged parts. Inspect condition of crank and cam sensors. Cracked or severely oil soaked components should be replaced.

Task E.7   Inspect and diagnose exhaust system; determine needed repairs.

Exhaust manifolds can be made of cast iron or sheet metal. Sheet metal manifolds are usually made of stainless steel. Inspect exhaust manifolds for cracks and leaks. On vehicles with computer controlled fuel delivery systems, air entering the exhaust system through a crack or leak ahead of the oxygen sensor can cause driveability and emission control system problems. Replace a cracked manifold.

The exhaust manifold on carbureted and throttle body fuel injected (TBI) engines may be equipped with a manifold heat control valve. This valve is closed when the engine is cold to direct hot exhaust gases to the underside of the intake manifold, directly under the carburetor or TBI unit. The gases heat the manifold, improving fuel vaporization in the cold engine. The valve opens as the engine warms up and the added heat is not needed. If the manifold heat control valve is stuck open or fails to close when the engine is cold, the engine may stumble during acceleration. If the valve is stuck in the closed position, engine power will be reduced and the intake manifold will overheat. The floor of the intake manifold may crack.

With the engine cold and shut off, check to see if the valve moves freely. Older engines use a bimetal thermostatic spring to operate the valve. Grasp the valve counterweight and rotate it back and forth. In some applications the valve is opened and closed using a vacuum actuator. Connect a vacuum pump to the actuator and apply vacuum to test the valve. The valve shaft and bushings should be lubricated periodically with a special solvent that contains graphite. Check the vehicle manufacturer's service manual for lubricant information.

5 Sample Test for Practice

Sample Test

Please note the letter and number in parentheses following each question. They match the overview in section 4 that discusses the relevant subject matter. You may want to refer to the overview using this cross-referencing key to help with questions posing problems for you.

1. An engine that demonstrates a single cylinder misfire is being diagnosed performing a leak down test. Which of these is the Most-Likely cause of the condition?
   A. 5% leakage with air coming out of the crankcase.
   B. 10% leakage with air coming out of the crankcase.
   C. A faulty ignition wire.
   D. Retarded ignition timing. (A.9)

2. Which of the following steps is the technician LEAST-Likely to perform when pressing the wrist pin into the piston and connecting rods?
   A. Align the bores in the piston and connecting rod.
   B. Heat the small end of the rod.
   C. Make sure that position marks on the piston and connecting rod are oriented properly.
   D. Heat the wrist pin. (C.12)

3. Technician A says a special puller and installer tool is required to remove and install the vibration damper. Technician B says if the inertia ring on the vibration damper is loose, the damper must be replaced. Who is right?
   A. A only
   B. B only
   C. Both A and B
   D. Neither A nor B (C.13)

4. If new rings are installed without removing the ring ridge, which of these is the Most-Likely result?
   A. Piston skirt damage.
   B. Piston pin damage.
   C. Connecting rod bearing damage.
   D. Piston compression ring damage. (C.4 and C.12)

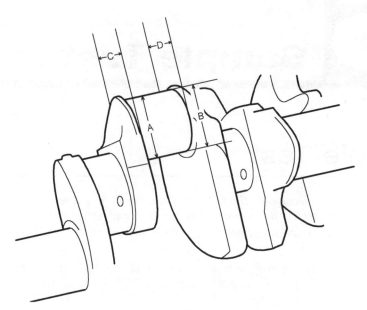

5. When measuring the crankshaft journal as shown in the figure, the difference between measurements:
   A. A and B indicates horizontal taper.
   B. C and D indicates vertical taper.
   C. A and C indicates out-of-round.
   D. A and D indicates out-of-round.   (C.5)

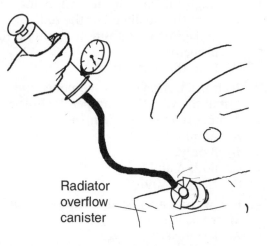

Radiator
overflow
canister

6. The tester in the figure may be used to test all the following items **EXCEPT:**
   A. cooling system leaks.
   B. the radiator cap pressure relief valve.
   C. coolant specific gravity.
   D. heater core leaks.   (D.3)

7. A loose belt may cause all of these **EXCEPT:**
   A. A discharged battery.
   B. Water pump bearing failure.
   C. Poor power steering assist.
   D. Engine overheating.   (D.4)

8. All of the following statements regarding manifold heat control valves are true
   EXCEPT:
   A. A manifold heat control valve improves fuel vaporization in the intake
      manifold especially when the engine is cold.
   B. A manifold heat control valve stuck in the closed position causes a loss of
      engine power.
   C. A manifold heat control valve stuck in the open position may cause an
      acceleration stumble.
   D. A manifold heat control valve stuck in the closed position reduces intake
      manifold temperature.   (E.8)

9. Which of these is the most common example of an internal oil cooler?
   A. Transmission cooler.
   B. Radiator.
   C. Engine oil cooler.
   D. Heater core.   (D.11)

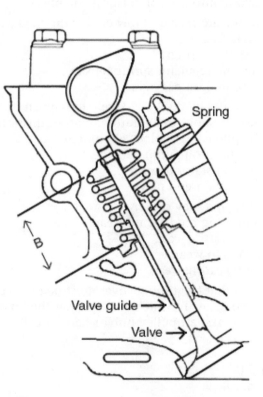

10. Measurement B in the figure is more than specified. Technician A says this
    problem may bottom the lifter plunger. Technician B says a shim could be
    installed under the valve spring. Who is right?
    A. A only
    B. B only
    C. Both A and B
    D. Neither A nor B   (B.9)

11. Technician A says that stuck valves may cause bent pushrods. Technician B says
    that improper valve timing may cause bent pushrods. Who is right?
    A. A only
    B. B only
    C. Both A and B
    D. Neither A nor B   (B.11)

12. All of the following oil pump measurements should be performed with a feeler gauge **EXCEPT:**
    A. Measure pump cover flatness with a feeler gauge positioned between a straightedge and the cover.
    B. Measure the clearance between the inner rotor and the housing.
    C. Measure the clearance between the inner and outer rotors with the rotors installed.
    D. Measure the clearance between the top of the rotors and a straightedge positioned across the top of the oil pump.   (D.2)

13. Which of these would be the LEAST-Likely to happen when bolting an OHC cylinder head on to a block with a warped deck surface?
    A. Premature cam bearing wear.
    B. Distorted valve seats.
    C. Main bearing failure.
    D. Coolant leakage into the combustion chambers.   (C.2)

14. All of the following are causes of low engine oil pressure **EXCEPT:**
    A. worn camshaft bearings.
    B. worn crankshaft bearings.
    C. weak oil pressure regulator spring tension.
    D. Restricted pushrod oil passages.

15. Which of these would be the Most-Likely to happen if the flywheel is bolted to a distorted crankshaft flange in an automatic-transmission-equipped vehicle?
    A. Improper bell housing to block alignment.
    B. Transmission front oil pump damage.
    C. Torque converter stator damage.
    D. Excessive crankshaft end play.   (C.14)

16. A bent connecting rod may cause:
    A. uneven connecting rod bearing wear.
    B. uneven main bearing wear.
    C. uneven piston pin wear.
    D. excessive cam bearing wear.   (C.11)

17. Technician A says that the timing cover must be removed in order to replace the crankshaft front oil seal. Technician B says that the lip on the front oil seal must face toward the crankshaft pulley hub or harmonic balancer. Who is right?
    A. A only
    B. B only
    C. Both A and B
    D. Neither A nor B   (C.16)

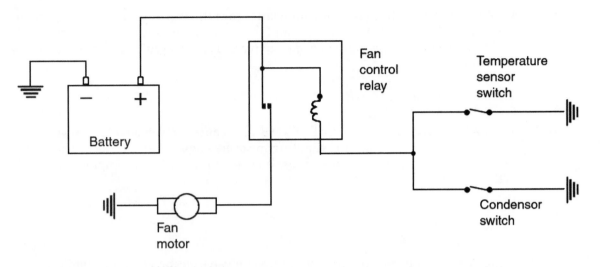

18. In the figure, an open ground circuit on the engine temperature sensor switch
    may cause:
    A. continual cooling fan motor operation.
    B. a completely inoperative cooling fan motor.
    C. a burned-out cooling fan motor.
    D. engine overheating.   (D.10)

19. Which of these is the LEAST-Likely place to use RTV sealer?
    A. Oil pan
    B. Valve cover.
    C. Timing cover.
    D. Cylinder head.   (C.16)

20. An upper radiator hose that collapses when engine RPM is raised with a fully
    warmed up engine may be caused by:
    A. Damaged radiator cap seal.
    B. An air leak at one of the bypass hoses.
    C. Water pump cavitation.
    D. A faulty upper radiator hose.   (D.9)

21. Which of these is the Most-Likely to happen if the radiator cap has a missing seal?
    A. Coolant recovery bottle will be empty.
    B. Engine will not reach normal operating temperature.
    C. Thermostat failure.
    D. Air will enter the cooling system.   (D.7)

22. Technician A says a defective water pump bearing may cause a growling noise
    when the engine is idling. Technician B says the water pump bearing may be
    ruined by coolant leaking past the pump seal. Who is right?
    A. A only
    B. B only
    C. Both A and B
    D. Neither A nor B   (D.8)

23. Technician A says worn valve stem seals may cause rapid valve stem and guide
    wear. Technician B says worn valve stem seals may cause excessive oil
    consumption. Who is right?
    A. A only
    B. B only
    C. Both A and B
    D. Neither A nor B   (B.5)

24. While measuring valve springs, Technician A says the valve spring must be rotated while measuring squareness. Technician B says that spring squareness can be checked by rolling the spring on a surface plate. Who is right?
    A. A only
    B. B only
    C. Both A and B
    D. Neither A nor B   (B.3)

25. Technician A says worn valve lock grooves may cause the valve locks to fly out of place with the engine running, resulting in severe engine damage. Technician B says worn valve lock grooves may cause a clicking noise with the engine idling. Who is right?
    A. A only
    B. B only
    C. Both A and B
    D. Neither A nor B   (B.4)

26. Which of these must be performed to measure valve clearance on mechanical valve trains?
    A. Back off each adjuster 2 turns.
    B. Rotate the cam to the lowest lift point for each cylinder.
    C. Move each cylinder to the beginning of the intake stroke.
    D. Adjust each valve to zero lash before taking measurements.   (B.12)

27. Which of these would be the best to use to measure crank end play?
    A. Micrometer.
    B. Snap gauge.
    C. Dial indicator.
    D. Feeler gauge.   (C.7)

28. The customer complains that the engine cranks but does not start; the first thing to check should be:
    A. valve train operation.
    B. battery voltage.
    C. compression.
    D. engine vacuum.   (A.2)

29. Technician A says that the coolant bypass hose allows coolant flow through the block while the thermostat is closed. Technician B says that the heater core will receive coolant from the engine before the thermostat is open. Who is right?
    A. A only
    B. B only
    C. Both A and B
    D. Neither A nor B   (D.6)

30. When discussing camshaft-bearing clearance, Technician A says excessive camshaft bearing clearance may result in lower-than-specified oil pressure. Technician B says excessive camshaft bearing clearance may cause a clicking noise when the engine is idling. Who is right?
    A. A only
    B. B only
    C. Both A and B
    D. Neither A nor B   (B.15)

31. On engines where the camshaft drive gear teeth mesh directly with the crankshaft gear teeth, Technician A says the timing gear backlash may be measured with a dial indicator. Technician B says timing gear backlash may be measured with a micrometer. Who is right?
    A. A only
    B. B only
    C. Both A and B
    D. Neither A nor B   (B.13)

32. During a cylinder balance test on an engine with electronic fuel injection, one cylinder provides very little rpm drop. Technician A says the ignition system may be misfiring on that cylinder. Technician B says the engine may have an intake manifold vacuum leak. Who is right?
    A. A only
    B. B only
    C. Both A and B
    D. Neither A nor B   (A.7)

33. Technician A says improper valve timing may cause reduced engine power. Technician B says improper valve timing may cause bent valves in some engines. Who is right?
    A. A only
    B. B only
    C. Both A and B
    D. Neither A nor B   (B.16)

34. Technician A says hydraulic valve lifter bottoms should be flat or concave. Technician B says a sticking lifter plunger may cause a burned exhaust valve. Who is right?
    A. A only
    B. B only
    C. Both A and B
    D. Neither A nor B   (B.11)

35. Which of these is the first thing to do when removing the timing belt?
    A. Remove the valve cover(s).
    B. Remove belt tensioner.
    C. Align all timing marks.
    D. Remove water pump.   (B.13)

36. A cylinder head for an OHC engine is being inspected. The feeler gauge measurement shown in the diagram is greater than the maximum specification. Technician A says that the head should be resurfaced and reinstalled. Technician B says that the camshaft bores should be measured as well. Who is right?
    A. A only
    B. B only
    C. Both A and B
    D. Neither A nor B   (B.2)

37. The hose from the positive crankcase ventilation (PCV) valve to the intake manifold is restricted. This problem could result in:
    A. an acceleration stumble.
    B. oil accumulation in the air cleaner.
    C. engine surging at high speed.
    D. engine detonation during acceleration.   (E.6)

38. In the figure, the technician is Most-Likely checking:
    A. valve guide depth.
    B. valve seat angle.
    C. cylinder head flatness.
    D. valve seat runout.   (B.9)

39. Technician A says that a soft or gummy heater hose may be caused by a missing exhaust manifold heat shield. Technician B says that a brittle or hard lower radiator hose may be caused by engine oil leaking onto the hose. Who is right?
    A. A only
    B. B only
    C. Both A and B
    D. Neither A nor B   (D.5)

40. Technician A says that when measuring piston ring groove to ring clearance, you should place a ring into the groove and measure clearance with a feeler gauge. Technician B says that the ring groove clearance should be the same as the ring end gap clearance. Who is right?
    A. A only
    B. B only
    C. Both A and B
    D. Neither A nor B   (C.12)

41. Reduced turbocharger boost pressure may be caused by a:
    A. wastegate valve stuck closed.
    B. wastegate valve stuck open.
    C. leaking wastegate diaphragm.
    D. disconnected wastegate linkage.   (E.3)

42. While discussing torque-to-yield head bolts, Technician A says compared to conventional head bolts, torque-to-yield bolts provide a more uniform clamping force. Technician B says torque-to-yield bolts are tightened to a specific torque and then rotated tighter a certain number of degrees. Who is right?
    A. A only
    B. B only
    C. Both A and B
    D. Neither A nor B   (B.18)

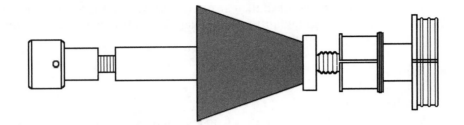

43. The tool shown in the figure is used to:
    A. remove camshaft bearings only.
    B. install camshaft bearings only.
    C. measure camshaft bearing alignment.
    D. both remove and install camshaft bearings.   (C.8)

44. Technician A says improper balance shaft timing causes severe engine vibrations.
    Technician B says the balance shafts are timed in relation to the camshaft. Who is
    right?
    A. A only
    B. B only
    C. Both A and B
    D. Neither A nor B   (C.9)

45. Technician A says that an important step in resolving a customer's concern is to
    test drive the vehicle to verify the concern. Technician B says that an important
    step in resolving a customer's concern is to check to see if the problem has an
    associated technical service bulletin.
    A. A only
    B. B only
    C. Both A and B
    D. Neither A nor B   (A.1)

46. With the engine idling, a vacuum gauge connected to the intake manifold
    fluctuates as shown in the figure. These vacuum gauge fluctuations may be caused
    by:
    A. late ignition timing.
    B. intake manifold vacuum leaks.
    C. a restricted exhaust system.
    D. sticky valve stems and guides.   (A.6)

47. A battery rated at 600 cold cranking amps (cca) is load tested at one-half of its
    rated cca for 15 seconds. The results show 10.1 volts. The results indicate that this
    battery:
    A. is satisfactory.
    B. needs recharging.
    C. is bad and should be replaced.
    D. should be retested at load for 30 seconds.   (E.4)

48. Oil is leaking from the crankshaft rear main bearing seal on an engine. Technician A says the oil seal could be faulty. Technician B says the PCV system may not be functioning. Who is right?
    A. A only
    B. B only
    C. Both A and B
    D. Neither A nor B   (A.3)

49. The customer describes a noise that is like a lawn mower engine during acceleration. Which of these might cause this type of noise?
    A. An intake manifold leak.
    B. The choke stuck closed.
    C. A restricted air filter.
    D. An exhaust manifold leak.   (E.7)

50. A loud thumping noise is present during all engine speeds. If the oil pressure is normal, which of these would be the Most-Likely cause?
    A. Worn pistons and cylinders.
    B. Loose flywheel bolts.
    C. Worn main bearings.
    D. Loose camshaft bearings.   (A.4)

6 Additional Test Questions for Practice

Additional Test Questions

Please note the letter and number in parentheses following each question. They match the overview in section 4 that discusses the relevant subject matter. You may want to refer to the overview using this cross-referencing key to help with questions posing problems for you.

1. Technician A says valve stem height is measured from the bottom of the valve guide to the top of the valve guide. Technician B says that excessive valve stem to guide clearance may result in excessive oil consumption. Who is right?
   A. A only
   B. B only
   C. Both A and B
   D. Neither A nor B   (B.6)

2. The LEAST-Likely cause of excessive blue smoke in the exhaust of a turbocharged engine is:
   A. a PCV valve stuck in the open position.
   B. worn turbocharger seals.
   C. worn valve guide seals.
   D. worn piston rings.   (E.3)

3. An engine is being disassembled after spinning a rod bearing. Technician A says to flush the oil passages in the head and block. Technician B says to replace or flush the oil cooler. Who is right?
   A. A only
   B. B only
   C. Both A and B
   D. Neither A nor B   (C.1)

4. All of the following could cause a bent pushrod **EXCEPT:**
   A. worn cam bearings.
   B. a broken timing chain.
   C. a sticking valve.
   D. improper valve adjustment.   (B.10)

5. Technician A says that room temperature vulcanizing (RTV) sealant is used to secure threaded fasteners. Technician B says that fumes from an anaerobic sealant can damage an oxygen sensor. Who is right?
   A. A only
   B. B only
   C. Both A and B
   D. Neither A nor B   (C.16)

6. On an overhead cam (OHC) cylinder head with removable bearing caps, which of the following is used to measure bearing alignment?
   A. A straightedge
   B. Plastigauge
   C. A dial indicator
   D. A telescoping gauge   (B.16)

7. Technician A says that balance shafts should be checked for runout following the same procedure used for measuring camshaft runout. Technician B says that balance shaft journals should be measured for taper following the same procedure used for measuring crankshaft journal taper. Who is right?
   A. A only
   B. B only
   C. Both A and B
   D. Neither A nor B   (C.9)

8. Technician A says water pump noise may be caused by faulty shaft bearings. Technician B says noise may be caused by the water pump impeller contacting the timing cover. Who is right?
   A. A only
   B. B only
   C. Both A and B
   D. Neither A nor B   (D.8)

9. An electric cooling fan is inoperative. Technician A says this could be caused by a bad ground in the cooling fan circuit. Technician B says this could be caused by a bad wire to the fan relay. Who is right?
   A. A only
   B. B only
   C. Both A and B
   D. Neither A nor B   (D.10)

10. A technician is servicing an overhead camshaft (OHC) cylinder head where the camshaft runs without bearings in its bore. If camshaft journal-to-bore clearance exceeds specification, the technician must:
    A. replace bearings with oversized bearings.
    B. replace the camshaft.
    C. insert bushings.
    D. replace the cylinder head.   (B.13)

11. Technician A says piston ring grooves should be cleaned by using a file. Technician B says to position a feeler gauge between each ring and the ring groove to measure the ring groove clearance. Who is right?
    A. A only
    B. B only
    C. Both A and B
    D. Neither A nor B   (C.10)

12. Technician A says that replacing a 180° F (82° C) thermostat with a 195° F (91° C) thermostat will cause the engine to warm up faster. Technician B says that removing the thermostat from an engine may cause "hot spots" to develop in the engine. Who is right?
    A. A only
    B. B only
    C. Both A and B
    D. Neither A nor B   (D.6)

13. If either the radiator pressure cap sealing gasket or the radiator filler neck seat are damaged, which of the following is LEAST-Likely to occur?
    A. The lower radiator hose will burst.
    B. The engine coolant will boil.
    C. The engine coolant will overflow.
    D. The engine will overheat.   (D.9)

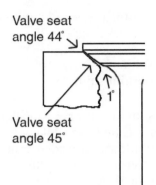

Valve seat
angle 44°

1°

Valve seat
angle 45°

14. In the figure, Technician A says that a three-angle valve job is shown. Technician B
says that poor valve face to valve seat orientation is shown. Who is right?
    A. A only
    B. B only
    C. Both A and B
    D. Neither A nor B   (B.7)

15. After a vehicle is parked overnight and then started in the morning, the engine
has a lifter noise that disappears after running for a short while. The Most-Likely
cause would be:
    A. low oil pressure.
    B. low oil level.
    C. a worn lifter bottom.
    D. excessive lifter leak-down.   (A.4)

16. When using a compression tester, as shown, the compression readings on the
cylinders are all even, but lower than the specified compression. This could
indicate:
    A. a blown head gasket.
    B. carbon buildup.
    C. a cracked head.
    D. worn rings and cylinders.   (A.8)

17. Technician A says that an overtensioned V-belt can damage the alternator front
bearing. Technician B says that an overtensioned V-belt can cause the upper half
of the crankshaft front main bearing to wear prematurely. Who is right?
    A. A only
    B. B only
    C. Both A and B
    D. Neither A nor B   (D.4)

18. Valve spring installed height is measured between the lower edge of the top retainer and the:
    A. cylinder head.
    B. top edge of the top shim.
    C. bottom edge of the bottom shim.
    D. spring seat.   (B.10)

19. A technician is preparing to install new oil gallery plugs in a cylinder block. Which of these operations is he LEAST-Likely to perform?
    A. Run a bottoming tap into the threaded gallery holes.
    B. Apply oil resistant sealer to the new plugs.
    C. Apply teflon tape to the threaded plugs.
    D. Run a rifle brush through the galleries.   (C.3)

20. When checking the installed spring seat pressure a spring fails to meet the minimum specification and the installation of a spring shim, adequate to return seat pressure to spec results in coil bind at maximum valve lift. Which of these is the best solution?
    A. Replace the valve seat to correct the installed height.
    B. Replace the spring to correct seat pressure.
    C. Replace the valve to correct seat pressure.
    D. Use the next thinnest spring shim that will not cause coil bind.   (B.3)

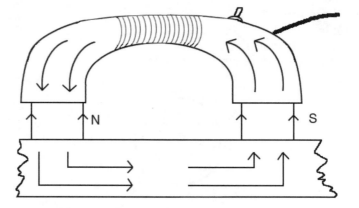

21. An electromagnetic-type tester, as shown in the figure, and iron filings may be used to check for cracks in:
    A. aluminum cylinder heads.
    B. pistons.
    C. cast-iron cylinder heads.
    D. aluminum intake manifolds.   (B.2)

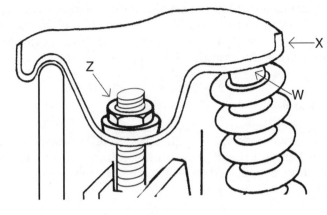

22. In the figure, how is valve lash adjusted?
    A. By adding shims to point W
    B. By adding shims to point X
    C. No adjustment is required
    D. By turning nut Z   (B.12)

23. Which of the following would LEAST-Likely require crankshaft grinding?
    A. Excessive taper
    B. An out-of-round journal
    C. Excessive journal scoring
    D. Excessive thrust wear   (C.5)

24. Technician A says that in the figure, X can be replaced without removing the head. Technician B says Y can be replaced without removing the head. Who is right?
    A. A only
    B. B only
    C. Both A and B
    D. Neither A nor B   (B.5)

25. A cylinder balance test is being performed on a engine to determine which cylinder is causing a "miss." Technician A says that when the faulty cylinder is disabled, engine rpm will drop more than for the other cylinders. Technician B says disabling the faulty cylinder will cause the engine to stall. Who is right?
    A. A only
    B. B only
    C. Both A and B
    D. Neither A nor B   (A.7)

26. All lifters in an overhead valve engine are cupped (concave). Technician A says that the camshaft and the lifters must be replaced. Technician B says that the rocker arms must be replaced to return correct lifter pre-load. Who is right?
    A. A only
    B. B only
    C. Both A and B
    D. Neither A nor B   (B.11)

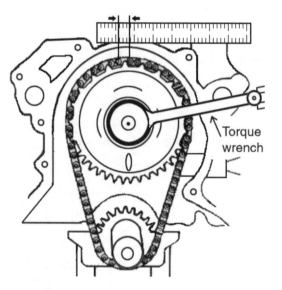

27. In the figure, what is being performed?
    A. Adjusting cam timing
    B. Locating TDC
    C. Measuring timing chain stretch
    D. Adjusting valve lash   (B.14)

28. A cast iron cylinder block has just been hot tanked and is ready for inspection. Technician A says that the block deck should be checked for warpage using a straightedge and a feeler gauge. Technician B says that minor nicks or burrs on the block deck can be removed using a whetstone or a file. Who is right?
    A. A only
    B. B only
    C. Both A and B
    D. Neither A nor B   (C.2)

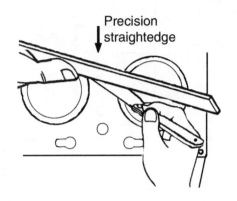

Precision straightedge

29. In the diagram shown what is being measured?
    A. Valve guide height
    B. Valve seat depth
    C. Cylinder head warpage
    D. Installed height   (B.2)

30. After performing a compression test on a V-8 engine, two cylinders have pressure readings of 60 psi (44 kPa) while the others have a reading of 135 psi (931 kPa). The two low cylinders are next to each other. Technician A says this could be caused by a loose timing chain. Technician B says a leaking head gasket could cause this. Who is right?
    A. A only
    B. B only
    C. Both A and B
    D. Neither A nor B   (A.8)

31. An engine is idling at 750 rpm. The pointer on the vacuum gauge in the figure is floating between 11 and 16 in. Hg. The Most-Likely cause would be:
    A. retarded timing.
    B. advanced timing.
    C. a stuck EGR valve.
    D. too lean an idle mixture.   (A.6)

32. A defective water pump can be diagnosed by all of the following **EXCEPT** by:
    A. observing residue at the water pump drain hole.
    B. observing a coolant leak from the water pump.
    C. hearing a groaning noise at cruising speeds.
    D. using a pressure tester.   (D.8)

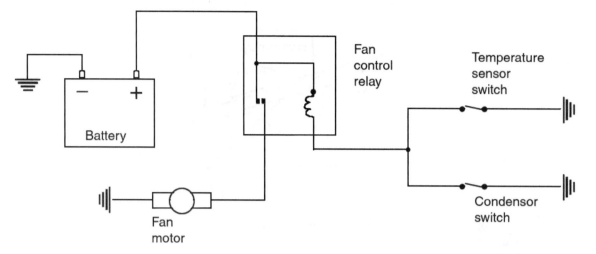

33. An electric drive cooling fan circuit is shown. Technician A says if the coolant temperature sensor switch is stuck closed, the cooling fan will stop when the ignition is turned off. Technician B says when the air conditioner (A/C) is turned on, the fan relay winding will be grounded through the condenser switch to activate the cooling fan. Who is right?
    A. A only
    B. B only
    C. Both A and B
    D. Neither A nor B   (D.10)

34. Technician A says that some head bolts stretch permanently when they are tightened. Technician B says some head bolts cannot be reused. Who is right?
    A. A only
    B. B only
    C. Both A and B
    D. Neither A nor B   (B.18)

35. Technician A says that stretched main bearing bores can be corrected by filing the main bearing caps. Technician B says this problem can be corrected by replacing the main bearing caps. Who is right?
    A. A only
    B. B only
    C. Both A and B
    D. Neither A nor B   (C.6)

36. When checking connecting rods for damage and wear, which of the following is LEAST-Likely to be checked?
    A. Rod center-to-center length
    B. Rod straightness
    C. Small end bore condition
    D. Big end bore out-of-round   (C.11)

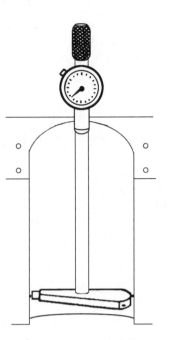

37. The gauge in the figure is being used to check cylinder diameter near the bottom of the cylinder. If the technician wants to determine cylinder taper, where must he take an additional measurement?
    A. At the top of the cylinder, just above the ring ridge
    B. At the top of the cylinder, just below the ring ridge
    C. Near the top of the cylinder, at the top of the oil ring contact area
    D. Near the bottom of the cylinder, at 90° to the first measurement   (C.4)

38. Technician A says that compression rings should never be installed by "spiraling" them onto the piston. Technician B says that the top compression ring should always be installed on the piston first. Who is right?
    A. A only
    B. B only
    C. Both A and B
    D. Neither A nor B   (C.12)

39. The rocker arms on a pushrod engine have a 1.5:1 ratio. This means that:
    A. a cam lift of 0.250 inch (6.35 mm) will cause the valve to open 0.188 inch (4.76 mm).
    B. a cam lift of 0.250 inch (6.35 mm) will cause the valve to open 0.375 inch (9.53 mm).
    C. the engine must be fitted with hydraulic lifters.
    D. the engine must be fitted with roller lifters.   (B.10)

40. Valve face and seat concentricity can be measured with all of the following **EXCEPT:**
    A. a concentricity tester.
    B. a dial caliper.
    C. blue dye.
    D. a dial indicator.   (B.8)

41. The areas around the mounting holes on a sheet metal rocker cover are dished from having the fasteners overtightened. To prevent oil leaks from occurring when the rocker cover is installed, the technician should:
    A. replace the rocker cover with a new one.
    B. hammer the dished areas flat again.
    C. use two gaskets instead of one.
    D. use RTV sealant instead of a gasket.   (C.15)

42. Technician A says that valve lock grooves on the valve stems must be inspected for rounded shoulders. Technician B says valve stems having rounded or uneven shoulders require machining. Who is right?
    A. A only
    B. B only
    C. Both A and B
    D. Neither A nor B   (B.4)

43. Most manufacturers recommend that piston diameter be measured at 90° to the wrist pin bore:
    A. at the very top of the piston.
    B. at the wrist pin bore centerline.
    C. about 3⁄4 inch below the wrist pin bore centerline.
    D. about 1⁄4 inch from the bottom of the piston skirt.   (C.10)

44. Vibration damper rubber should be inspected for all of the following **EXCEPT:**
    A. hub contact area scoring.
    B. looseness.
    C. cracks.
    D. oil soaking.   (C.13)

45. The figure shows the cylinder and ring ridge. If the amount of cylinder wear does not require cylinder reboring, Technician A says the ring ridge at the top of the each cylinder can be removed with 400-grit sand paper. Technician B says the ring ridge at the top of the each cylinder can be removed with a 200-grit bead hone. Who is right?
    A. A only
    B. B only
    C. Both A and B
    D. Neither A nor B   (C.4)

46. While examining the old connecting rod bearings from an engine, the technician notices that the bearings from one rod are worn along the parting lines. This means that the technician should check carefully for:
    A. rod stretch.
    B. rod twisting.
    C. rod bending.
    D. a loose wrist pin.   (C.11)

47. When installing a piston/connecting rod into the cylinder block, which of the following steps is a technician LEAST-Likely to perform?
   A. Position the crankshaft journal at bottom dead center (BDC).
   B. Install boots over the rod bolts.
   C. Make sure the rings are installed right side up.
   D. Check piston-to-cylinder wall clearance using Plastigauge.   (C.12)

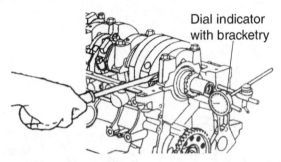

Dial indicator with bracketry

48. In the figure shown, Technician A says noise may be present if the measurement is too large. Technician B says that main bearing damage will occur if the measurement is too small. Who is right?
   A. A only
   B. B only
   C. Both A and B
   D. Neither A nor B   (C.5)

49. The procedure for aligning cam and crankshaft sprockets before installing a timing chain varies from manufacturer to manufacturer. One step common to most procedures, however, is for the technician to:
   A. rotate the camshaft to fully open the intake valve in cylinder number 1.
   B. rotate the camshaft to fully open the exhaust valve in cylinder number 1.
   C. rotate the crankshaft to position piston number 1 at TDC.
   D. rotate the crankshaft to position piston number 1 at BDC.   (B.16)

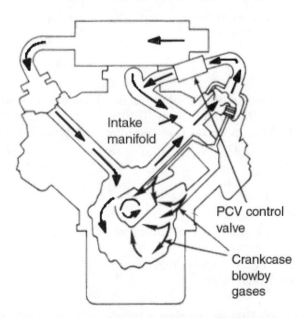

Intake manifold

PCV control valve

Crankcase blowby gases

50. The figure shows the positive crankcase ventilation (PCV) system. All of the following are symptoms of a stuck open PCV valve **EXCEPT:**
   A. blowby gases in the air filter.
   B. the engine stalling.
   C. rough idle operation.
   D. a lean air/fuel ratio.   (E.5)

51. The oil light on a vehicle stays on while the engine is running. Technician A say this could be caused by too much cam bearing clearance. Technician B says a grounded wire in the oil lamp warning lamp circuit could cause this. Who is right?
    A. A only
    B. B only
    C. Both A and B
    D. Neither A nor B   (D.1)

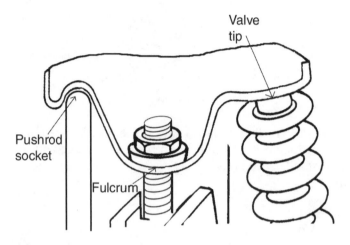

52. The figure shows an example of the rocker arm assembly on an engine equipped with hydraulic lifters. When adjusting valve lash on this engine while it is running, the step LEAST-Likely to be performed by a technician is:
    A. turning the adjusting nut clockwise ¼ turn at a time.
    B. turning the adjusting nut clockwise 2 turns at a time.
    C. turning the adjusting nut counterclockwise until a clicking noise occurs.
    D. installing oil shrouds on the rocker arm.   (B.12)

53. All of the following are reasons to replace a hydraulic valve lifter **EXCEPT:**
    A. excessive bleed-down.
    B. convex bottom.
    C. pitted bottom.
    D. flat bottom.   (B.11)

54. A vehicle is equipped with a coolant recovery system. Coolant does not return to the radiator when the engine cools. Technician A says that the transfer hose may be plugged. Technician B says that the filler neck soldered joint could be cracked. Who is right?
    A. A only
    B. B only
    C. Both A and B
    D. Neither A nor B   (D.9)

55. The LEAST-Likely cause of an oil saturated PCV filter is:
    A. worn piston rings.
    B. an obstructed PCV vacuum hose.
    C. a stuck open PCV valve.
    D. a clogged PCV valve.   (E.5)

56. The customer says that the engine requires excessive cranking to start. The LEAST-Likely cause of this problem would be:
    A. a cracked cylinder block.
    B. a jumped timing belt.
    C. a faulty fuel pump.
    D. a stuck-open EGR valve.   (A.2)

57. A engine equipped with electronic fuel injection has a loose exhaust manifold. Technician A says that the loose manifold may cause noisy engine operation. Technician B says that the loose manifold may cause poor vehicle driveability. Who is right?
    A. A only
    B. B only
    C. Both A and B
    D. Neither A nor B   (E.7)

58. Serpentine belt stretch is indicated by:
    A. using a belt tension gauge.
    B. belt deflection.
    C. a squealing noise at idle.
    D. using the scale on the tensioner housing.   (D.4)

59. Which of these is NOT true of torque to yield attaching hardware?
    A. They must always be discarded after use.
    B. Provide more even clamping.
    C. They require special tightening procedures.
    D. Are occasionally used for connecting rod applications.   (B.17)

60. Technician A says that an engine oil cooler can be located inside one of the radiator tanks. Technician B says that an engine oil cooler can be mounted ahead of the radiator support. Who is right?
    A. A only
    B. B only
    C. Both A and B
    D. Neither A nor B   (D.11)

61. A technician is testing an upper radiator hose by squeezing it. The most likely cause of crackling or crunching noises would be:
    A. a corroded anti-collapse spring.
    B. low coolant level.
    C. a deteriorated hose inner liner.
    D. damage due to contact with power steering fluid.   (D.5)

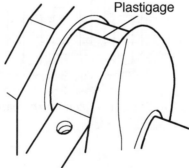

62. To measure bearing clearance, install a strip of Plastigage across the journal, as in the figure, and then tighten the bearing cap to the specified torque. Remove the bearing cap and measure the width of the Plastigage on the journal with which of the following?
    A. The Plastigage package.
    B. A ruler.
    C. Dial calipers.
    D. A micrometer.   (C.7)

63. A technician is using a dial indicator to measure valve-to-guide clearance. When the valve head is moved from side to side, the dial indicator shows a maximum value of 0.004 inch (0.102 mm). This means that valve guide clearance is:
    A. 0.001 inch (0.025 mm).
    B. 0.002 inch (0.051 mm).
    C. 0.004 inch (0.102 mm).
    D. 0.008 inch (0.203 mm).   (B.6)

64. During a cylinder balance test on a port fuel injected engine with coil-on-plug ignition, one cylinder is found to have virtually no rpm change. Which of these is the Most-Likely cause?
    A. A faulty crank position sensor.
    B. A faulty fuel injector.
    C. A vacuum leak at the throttle body.
    D. A fuel saturated vapor canister.   (A.7)

65. Technician A says that valve rotators should be removed and cleaned during an engine overhaul. Technician B says that a rotator causing the valve to rotate in either direction is functioning properly.
    A. A only
    B. B only
    C. Both A and B
    D. Neither A nor B   (B.4)

66. Technician A says that a press-fit harmonic balancer should be removed using a three-jaw gear puller. Technician B says that a harmonic balancer with a damaged keyway should be replaced. Who is right?
    A. A only
    B. B only
    C. Both A and B
    D. Neither A nor B   (C.13)

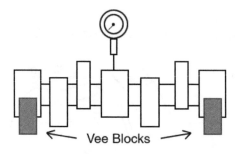

Vee Blocks

67. The measuring tool in the figure is checking the camshaft:
    A. journal condition.
    B. runout.
    C. lift.
    D. bearing clearance.   (B.15)

68. An excessive sulfur smell in the exhaust of a vehicle with a catalytic converter can be an indication of:
    A. a lean fuel mixture.
    B. coolant leaking into a combustion chamber.
    C. a rich fuel mixture.
    D. a vacuum leak.   (A.5)

69. A low, steady vacuum gauge reading as shown above indicates:
    A. burned or leaking valves.
    B. late ignition timing.
    C. weak valve springs.
    D. a leaking head gasket. (A.6)

70. During a cylinder leakage test one cylinder is found to have 90% leakage on two cylinders next to one another. Neither cylinder will stay at top dead center during the test. Which of these is the Most-Likely cause?
    A. Burned exhaust valves on both cylinders.
    B. Holes in the top of the pistons.
    C. A damaged head gasket.
    D. A collapsed lifters or lash adjusters both cylinders. (A.9)

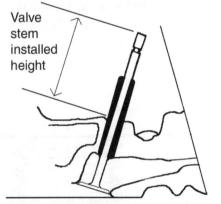

Valve stem installed height

71. In the above figure, Technician A says the tool is being used to check combustion chamber volume. Technician B says the tool is being used to check valve stem installed height. Who is right?
    A. A only
    B. B only
    C. Both A and B
    D. Neither A nor B (B.9)

72. During a cylinder leakage test, air comes out the PCV valve opening in the rocker arm cover. This is an indication of:
    A. worn intake valves.
    B. worn exhaust valves.
    C. a broken PCV valve.
    D. worn piston rings. (A.9)

73. Engines that use a hydraulic timing chain tensioner can be checked for chain stretch by:
    A. Checking the chain for side deflection.
    B. Measuring oil pressure at the tensioner feed.
    C. Measuring tensioner actuator length.
    D. Rotating the crankshaft in both directions 45 degrees to determine slack.
    (B.13)

74. Technician A says that when removing a cylinder head from an overhead camshaft (OHC) engine, the timing belt or chain will have to be removed from the block. Technician B says if the timing belt is to be reused, mark its direction of rotation and match it during reassembly. Who is right?
    A. A only
    B. B only
    C. Both A and B
    D. Neither A nor B   (B.13)

75. Which of these should be performed first when a starter fails to crank?
    A. Measure static battery voltage.
    B. Remove spark plugs.
    C. Check for the presence of spark.
    D. Bypass starter solenoid with remote starter button.   (A.2)

76. Technician A says blue-gray smoke coming from the exhaust may be caused by carboned oil. Technician B says this could be caused by a plugged oil drain passage in the cylinder head. Who is right?
    A. A only
    B. B only
    C. Both A and B
    D. Neither A nor B   (A.5)

77. Technician A says that positive type valve stem seals must be installed before the valves are installed in the cylinder head. Technician B says that positive type valve stem seals must be pushed down firmly over the top of the valve guides. Who is right?
    A. A only
    B. B only
    C. Both A and B
    D. Neither A nor B   (B.5)

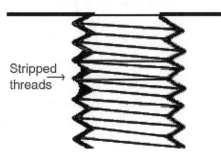

78. If threads are damaged, as shown in the figure, the opening may be drilled and threaded, and then a heli-coil may be installed to provide a thread:
    A. the same size as the new hole.
    B. one thread size smaller than the original.
    C. one thread size larger than the original.
    D. the same size as the original.   (C.3)

79. A cooling system is being checked with a pressure tester. The gauge rapidly rises when the engine is started. Technician A says this could be caused by a crack in the combustion chamber. Technician B says this could be caused by clogged tubes in the radiator. Who is right?
    A. A only
    B. B only
    C. Both A and B
    D. Neither A nor B   (D.3)

80. Technician A says that when installing an intake manifold that uses synthetic rubber seals at the front and rear ends, the top and bottom of the rubber seals should be coated with silicone sealer. Technician B says that only a dab of silicone sealer should be placed at the very ends of the seals. Who is right?
    A. A only
    B. B only
    C. Both A and B
    D. Neither A nor B   (E.1)

81. The LEAST-Likely cause of camshaft bind would be:
    A. excessive runout.
    B. an improperly installed bearing.
    C. bore misalignment.
    D. excessive bearing clearance.   (B.15)

82. A technician has added fluorescent dye to an engine crankcase in order to locate an oil leak. The dye will glow when it is exposed to:
    A. a fluorescent light.
    B. an ultraviolet light.
    C. a strobe light.
    D. an infrared light.   (A.3)

83. When installing a timing chain cover, the step that a technician is LEAST-Likely to perform would be:
    A. making sure the woodruff key is in place.
    B. making sure the oil slinger is in place.
    C. making sure the piston in cylinder #1 is at TDC.
    D. making sure the oil seal has been lubricated.   (C.16)

84. Technician A says that a damaged starter ring gear on a manual transmission flywheel can usually be replaced. Technician B says that a damaged starter ring gear on an automatic transmission flywheel can usually be replaced. Who is right?
    A. A only
    B. B only
    C. Both A and B
    D. Neither A nor B   (C.14)

85. While inspecting the intake manifold from a V-type engine, a technician notices a crack in the exhaust gas crossover passage on the underside of the manifold. The Most-Likely cause of this condition is:
    A. the intake manifold bolts were overtorqued.
    B. the intake manifold bolts were not torqued in the correct sequence.
    C. the EGR passage in the manifold is plugged with carbon.
    D. the heat riser valve on one of the exhaust manifolds is stuck shut.   (E.1)

86. Technician A says that camshaft lobe lift can be checked with the camshaft still mounted in the engine. Technician B says that camshaft runout can be checked with the camshaft still mounted in the engine. Who is right?
    A. A only
    B. B only
    C. Both A and B
    D. Neither A nor B   (B.15)

87. While replacing a heater hose, the technician discovers that the hose is stuck on the hose nipple. Which of these is the best solution to the problem?
    A. Run a screw driver around the inside of the hose and pry it off.
    B. Use pliers to twist and loosen the hose.
    C. Use a pick to loosen the hose and pry it off.
    D. Make a longitudinal slice in the hose and peel it from the nipple.   (D.5)

Appendices

Answers to the Test Questions for the Sample Test Section 5

1.	C	14.	D	27.	C	39.	D
2.	D	15.	B	28.	A	40.	A
3.	C	16.	A	29.	C	41.	B
4.	D	17.	D	30.	A	42.	C
5.	C	18.	D	31.	A	43.	D
6.	C	19.	D	32.	C	44.	A
7.	B	20.	D	33.	C	45.	C
8.	D	21.	D	34.	B	46.	D
9.	A	22.	C	35.	C	47.	A
10.	C	23.	B	36.	B	48.	C
11.	C	24.	A	37.	B	49.	D
12.	B	25.	A	38.	D	50.	B
13.	C	26.	B				

Explanations to the Answers for the Sample Test Section 5

Question #1
Answer A is wrong. 5% leakage is acceptable and will not cause a misfire.
Answer B is wrong. 10% leakage is acceptable and will not cause a misfire.
Answer C is correct. After eliminating cylinder loss the only possible choice in this list is a faulty ignition wire.
Answer D is wrong. Retarded timing will not cause a misfire. It would effect all cylinders if the engine could stay running with the timing retarded far enough to cause misfire.

Question #2
Answer A is wrong. The bores in the piston and the rod should be aligned before pressing in the wrist pin.
Answer B is wrong. The small end of the rod should be heated before pressing in the wrist pin.
Answer C is wrong. Position marks on the piston and rod should be oriented properly before pressing in the wrist pin.
Answer D is correct. Heating the wrist pin is a step LEAST-Likely performed by the technician. Heating the wrist pin will cause it to expand and prevent it from fitting into the connecting rod and piston bores.

Question #3
Answer A is wrong.
Answer B is wrong.
Answer C is correct. Technician A is correct in stating that a special puller and installer tool are required to remove and install the vibration damper. Using a regular gear puller to remove the vibration damper will damage the damper. Technician B is correct in stating that a loose inertia ring on the damper requires replacement of the damper.
Answer D is wrong.

Question #4
Answer A is wrong. The piston skirt would not be damaged.
Answer B is wrong. The piston pin is not affected by the ring ridge.
Answer C is wrong. The connecting rod bearings would be unaffected.
Answer D is correct. Failure to remove the ring ridge may cause piston ring land and/or top compression ring damage after the engine is assembled and started.

Question #5
Answer A is wrong. Measurements A and B indicate vertical taper.
Answer B is wrong. Measurements C and D indicate horizontal taper.
Answer C is correct. Referring to the illustration in question 6, the difference in measurement between A and C indicates out-of-round.
Answer D is wrong. Measurements A and D do not indicate out-of-round.

Question #6
Answer A is wrong. The tester may be used to test for cooling system leaks.
Answer B is wrong. The tester may be used to test the radiator cap pressure release valve.
Answer C is correct. Referring to the illustration in Question 6, this tester may be used to test for cooling system leaks, radiator cap pressure relief valve, and heater core leaks. This tester is not used to test for coolant specific gravity.
Answer D is wrong. The tester may be used to test for heater core leaks.

Question #7

Answer A is wrong. A loose belt may cause the battery to become discharged due to inadequate alternator output.

Answer B is correct. Water pump bearings will not fail due to a loose belt.

Answer C is wrong. Power steering assist might be diminished.

Answer D is wrong. If the belt that drives the water pump is slipping, the water pump may fail to move coolant effectively which could cause the engine to overheat.

Question #8

Answer A is wrong. A manifold heat control valve improves fuel vaporization in the intake manifold.

Answer B is wrong. A manifold heat control valve stuck in the closed position causes a loss of engine power.

Answer C is wrong. A manifold heat control valve stuck in the open position may cause acceleration stumble.

Answer D is correct. The exhaust manifold on carbureted and throttle body fuel-injected (TBI) engines may be equipped with a manifold heat control valve. If the manifold d heat control valve is stuck open or fails to close when the engine is cold, the engine may stumble during acceleration. If the valve is stuck in the closed position, engine power will be reduced and the intake manifold will overheat.

Question #9

Answer A is correct. The transmission cooler located in the radiator tank is the most common example of an internal cooler. Keep in mind that internal coolers are located inside another cooler and use the other coolers cooling medium to perform heat exchange. This makes almost all internal heat exchangers part of the radiator.

Answer B is wrong. The radiator is an external air to liquid heat exchanger.

Answer C is wrong. Engine oil coolers are almost always external air to liquid heat exchangers.

Answer D is wrong. A heater core is an external heat exchanger even though it is located inside the vehicle.

Question #10

Answer A is wrong.

Answer B is wrong.

Answer C is correct. Technician A's statement is true in non-adjustable valve train applications where the valve spring height is too high due to excessive valve or seat machining. This can cause the lifter to plunge too far resulting in valves that do not return to seat properly. Technician B's statement is also true. In situations where the valve train can compensate for the change in valve spring height or where the amount of shim necessary to bring the height into spec does not over-extend the lifter or lash adjuster.

Answer D is wrong.

Question #11

Answer A is wrong.

Answer B is wrong.

Answer C is correct. Technician A's statement is correct. Stuck valves may cause bent pushrods. Technician B's statement is also correct. Improper valve timing may cause bent pushrods.

Answer D is wrong.

Question #12

Answer A is wrong. Pump cover flatness may be measured with a feeler gauge.

Answer B is correct. The following oil pump measurements should be performed with a feeler gauge: Measure pump cover flatness with a feeler gauge positioned between a straightedge and the cover. Measure the clearance between the outer rotor and the housing. Measure the clearance between the inner and outer rotors with the rotors installed. Measure the clearance between the top of the rotors and a straightedge positioned across the top of the oil pump.

Answer C is wrong. The clearance between the inner and outer rotors may be measured with a feeler gauge while they are still installed.

Answer D is wrong. The clearance between the top of the rotors and a straightedge may be measured with a feeler gauge.

Question #13
Answer A is wrong. When a head is bolted to a warped cylinder block it can cause the head to distort which will load the cam bearing surfaces, causing uneven wear.
Answer B is wrong. The valve seats can become distorted and in some cases fall out of the head.
Answer C is correct. While it is possible that a severely distorted block might be damaged clear down to the main bearings, it is very unlikely, making this the best answer. Always remember when answering ASE test questions that the BEST answer is what you are looking for. When it comes to anything automotive there is an exception to every rule, so go with the most common and likely not the exception.
Answer D is wrong. Coolant leakage could occur once the engine is running.

Question #14
Answer A is wrong. Worn camshaft bearings may be a cause for low oil pressure.
Answer B is wrong. Worn crankshaft bearings may be a cause for low oil pressure.
Answer C is wrong. Weak oil pressure regulator spring tension may be a cause for low oil pressure.
Answer D is correct. Restricted push rod oil passages will not cause low oil pressure.

Question #15
Answer A is wrong. The position of the bell housing and block will not be affected.
Answer B is correct. While this condition is really only common to crankshafts that have a flange that is not part of the rear main bearing surface, it does include a lot of the engines produced in the last 50 years. Debris, grease, or damaged threads at the back of the crank can cause flywheel runout in any application. It is important to know that the front pump of an automatic transmission or transaxle will be damaged by this problem. Warpage of the flywheel flange almost never happens in normal service. This is a condition to watch for with a new engine that may have been damaged in shipping or a crank that was dropped while it was out of the engine.
Answer C is wrong. The torque converter stator is an internal component of the torque converter. It may be damaged as a result of the front pump coming apart and sending debris through the converter, but it will not be the root cause of the transmission failure.
Answer D is wrong. Since the crankshaft thrust surface is located elsewhere on the crank, the endplay will not be affected by a damaged flange.

Question #16
Answer A is correct. A bent connecting rod may cause uneven connection rod bearing wear.
Answer B is wrong. Uneven main bearings are not a cause for a bent connecting rod.
Answer C is wrong. Uneven piston pin wear is not a cause for a bent connecting rod.
Answer D is wrong. Excessive cam bearing wear is not a cause for a bent connecting rod.

Question #17
Answer A is wrong. On some engines, the crankshaft front oil seal can be replaced without removing the timing cover.
Answer B is wrong. Lip type seals are installed with the lip of the seal facing the fluid being sealed.
Answer C is wrong. Both Technicians A and B are wrong.
Answer D is correct.

Question #18
Answer A is wrong. An open ground circuit on the engine temperature sensor switch will prevent the cooling fan from operating. This would not cause continual operation. This may cause engine overheating.
Answer B is wrong. The cooling fan would operate when the condenser switch is closed.
Answer C is wrong. The cooling fan motor will operate when the condenser switch is closed.
Answer D is correct. An open ground circuit on the engine temperature sensor switch will prevent the cooling fan from operating. This may cause engine overheating.

Question #19
Answer A is wrong. RTV is commonly used for oil pans.
Answer B is wrong. Valve covers are another common use for RTV Silicone Sealer.
Answer C is wrong.
Answer D is correct. Use of RTV to seal a head gasket will cause a failure. Head gaskets are usually coated with their own heat activated sealer and do not require any sealer.

Question #20
Answer A is wrong. If the cap seal is damaged or missing, the system will not develop pressure and would not be able to hold a vacuum either.
Answer B is wrong. This is the same condition as answer A just a different leak.
Answer C is wrong. Water pump cavitation is a rare condition in late model engines and is usually the result of coolant that does not have adequate anti-foaming agent in it. There are usually far more problems with the coolant than just foaming when cavitation occurs.
Answer D is correct. While other problems such as a restriction in the cooling system can be present when the upper hose collapses, it is Most-Likely that the hose is weak and cannot retain it's shape under normal pressure differentials that can occur as an engine accelerates.

Question #21
Answer A is wrong. If the cap seal is missing, the coolant will usually run out of the radiator as the system warms up and will not use the recovery tank for expansion and contraction.
Answer B is wrong. The engine will still reach operating temperature.
Answer C is wrong. The thermostat is an internal component of the cooling system and will not be affected.
Answer D is correct. Air will enter the cooling system, which can lead to overheating, coolant loss, and rapid generation of deposits in the system.

Question #22
Answer A is wrong.
Answer B is wrong.
Answer C is correct. A defective water pump bearing may cause a growling noise when the engine is idling, and the water pump bearing may be ruined by coolant leaking past the pump seal.
Answer D is wrong.

Question #23
Answer A is wrong. Worn valve stem seals generally do not cause stem or guide wear.
Answer B is correct.
Answer C is wrong. Only Technician B is right.
Answer D is wrong. Technician B is right.

Question #24
Answer A is correct.
Answer B is wrong. Valve spring squareness is not checked by rolling the spring on a surface plate.
Answer C is wrong. Only Technician A is right.
Answer D is wrong. Technician A is right.

Question #25
Answer A is correct.
Answer B is wrong. Worn valve lock grooves will not cause a clicking noise at idle.
Answer C is wrong. Only Technician A is right.
Answer D is wrong. Technician A is right.

Question #26
Answer A is wrong.
Answer B is correct. To get correct measurements, each of the valves must be completely closed and the cam on the base circle (the bottom of the lobe).
Answer C is wrong.
Answer D is wrong.

Question #27
Answer A is wrong. A micrometer is used to measure thickness or width of an object.
Answer B is wrong. A snap gauge is used to take an internal measurement, as in the case of a cylinder.
Answer C is correct. The dial indicator is the best choice to measure the fore and aft clearance of the crankshaft thrust bearings.
Answer D is wrong. Using a feeler gauge to measure the crank endplay could damage the thrust bearings.

Question #28
Answer A is correct.
Answer B is wrong. If the engine is cranking properly, the battery would have sufficient voltage.
Answer C is wrong. Compression would not be the first test to perform.
Answer D is wrong. Because engine vacuum is low during cranking, a vacuum test would not be conclusive.

Question #29
Answer A is wrong.
Answer B is wrong.
Answer C is correct. The coolant bypass hose(s) allow coolant to flow through the entire engine and the heating system during warm up. When the thermostat is closed, the water pump draws coolant through the bypass hoses (in most applications) instead of the lower hose. The radiator is bypassed during warm-up to get the engine quickly and evenly up to operating temperature. The thermostat opens and allows coolant to flow to the radiator to regulate temperature.
Answer D is wrong.

Question #30
Answer A is correct.
Answer B is wrong. Excessive camshaft bearing clearance will not cause a clicking noise at idle.
Answer C is wrong. Only Technician A is right.
Answer D is wrong. Technician A is right.

Question #31
Answer A is correct.
Answer B is wrong. Camshaft gear backlash cannot be measured with a micrometer.
Answer C is wrong. Only Technician A is right.
Answer D is wrong. Technician A is right.

Question #32
Answer A is wrong.
Answer B is wrong.
Answer C is correct. Both a misfiring ignition system and an intake manifold leak could cause a cylinder to contribute too little power.
Answer D is wrong.

Question #33
Answer A is wrong.
Answer B is wrong.
Answer C is correct. Improper valve timing may cause reduced engine power and, on some engines, bent valves.
Answer D is wrong.

Question #34
Answer A is wrong. Lifter bottoms must be convex.
Answer B is correct.
Answer C is wrong. Only Technician B is right.
Answer D is wrong. Technician B is right.

Question #35
Answer A is wrong.
Answer B is wrong.
Answer C is correct. If the engine is an interference engine, this is absolutely critical. Even if the engine is not an interference engine, you will save yourself a lot of time by making sure that you align all of the marks in the way the manufacturer recommends.
Answer D is wrong.

Question #36
Answer A is wrong. Technician A's statement would be OK if he had suggested checking the camshaft saddles (bores) for warpage before machining the head.
Answer B is correct. Only Technician B's statement is totally accurate. This is another example of a type of Tech A/Tech B question that trip up many technicians when they are taking an ASE Test. If one statement is an absolute but it leaves out a critical item it is wrong. In this case Technician A described a complete operation. Technician B made a stand-alone statement that indicated the cam saddles should be checked. When you are taking an ASE test and you feel that the two technicians are arguing, be sure to re-read the question. ASE will not release a question where the two technicians are arguing.
Answer C is wrong.
Answer D is wrong.

Question #37
Answer A is wrong. A PCV valve stuck open may cause acceleration stumble.
Answer B is correct.
Answer C is wrong. A restricted PCV hose would not cause the engine to surge at high speed.
Answer D is wrong. A restricted PCV hose would not lead to engine detonation.

Question #38
Answer A is wrong. The tool shown does not measure valve guide depth.
Answer B is wrong. The tool shown does not measure valve seat angle.
Answer C is wrong. A straightedge and feeler gauge are used to measure cylinder head flatness.
Answer D is correct.

Question #39
Answer A is wrong. Excessive heat will cause a heater hose to become hard and brittle.
Answer B is wrong. Engine oil will cause a radiator hose to become soft and gummy.
Answer C is wrong. Both Technicians A and B are wrong.
Answer D is correct.

Question #40
Answer A is correct. Technician B's statement is wrong because ring end-gap and ring groove clearance have no relationship to one another.
Answer B is wrong.
Answer C is wrong.
Answer D is wrong.

Question #41
Answer A is wrong. A wastegate valve stuck closed would cause increased boost pressure.
Answer B is correct.
Answer C is wrong. A leaking wastegate diaphragm may cause increased boost pressure.
Answer D is wrong. A disconnected wastegate linkage would cause increased boost pressure because the valve would never open.

Question #42
Answer A is wrong.
Answer B is wrong.
Answer C is correct. Torque-to-yield bolts provide a more uniform clamping force than conventional bolts. They are often tightened by being torqued to a specific lb-ft value, and then rotated a specific number of degrees.
Answer D is wrong.

Question #43
Answer A is wrong. The tool can also be used to install camshaft bearings.
Answer B is wrong. The tool can also be used to remove camshaft bearings.
Answer C is wrong. The tool is not used for taking measurements.
Answer D is correct.

Question #44
Answer A is correct.
Answer B is wrong. Balance shafts should be timed in relation to the crankshaft.
Answer C is wrong. Only Technician A is right.
Answer D is wrong. Technician A is right.

Question #45
Answer A is wrong.
Answer B is wrong.
Answer C is correct. Checking for technical service bulletins can save time when solving a customer's concern. If you do not drive the vehicle to get an idea what the concern is, it will be difficult to diagnose it with confidence.
Answer D is wrong.

Question #46
Answer A is wrong. Late ignition timing would result in a low, steady reading.
Answer B is wrong. Intake manifold leaks would cause a very low, steady reading.
Answer C is wrong. A restricted exhaust system would cause vacuum to slowly decrease after the engine was accelerated and held steady.
Answer D is correct.

Question #47
Answer A is correct.
Answer B is wrong. The battery should recover the lost voltage over a short period of no demand.
Answer C is wrong. The battery passed the load test.
Answer D is wrong. The battery passed the load test.

Question #48
Answer A is wrong.
Answer B is wrong.
Answer C is correct. Oil leaking from the crankshaft rear main bearing seal could be caused by a faulty oil seal or a malfunctioning PCV system.
Answer D is wrong.

Question #49
Answer A is wrong. An intake manifold leak, if it can be heard, would be more of an air rushing or whistling noise.
Answer B is wrong. The choke itself does not make any noise. Watch for distracters on ASE tests that require you to "rationalize" that they work. They are not the answer.
Answer C is wrong. A really restricted air filter might make a whistle, but, again, it is not usually associated with any sound.
Answer D is correct. An exhaust manifold leak has a definite putt-putt sound to it that a customer might describe as a lawn mower sound.

Question #50

Answer A is wrong. The noise of piston wear is usually more like a screech. Sometimes when a piston skirt has excessive clearance you will hear a knock. Oil pressure will not be affected by this problem, so you must be careful to identify it correctly.

Answer B is correct. Here is the most common thumping noise that does not affect oil pressure. This is most prevalent when the bolts were not torqued properly during installation of the flywheel or flex-plate. The idea behind this question was to make you aware that external problems can sound like main bearing noise. It is important to use all of the symptoms available to narrow the problem down. This question is probably too ambiguous to survive ASE scrutiny, but it does serve its purpose to instruct here.

Answer C is wrong. In nearly all cases, worn main bearings will make noise and oil pressure will drop.

Answer D is wrong. Loose cam bearings are usually the result of something below them, like the crank or rod bearings coming apart. An engine with loose cam bearings was either assembled incorrectly or would be in a lot bigger trouble than the cam bearings making noise and dropping oil pressure. In some older high performance applications the spring pressure necessary to get big valves closed at high RPM could cause accelerated cam bearing wear. In overhead cam applications a timing belt installed too tight can cause damage to the cam bearings or saddles. ASE does not ask questions that assume a technician is incompetent or has performed a job incorrectly. There is enough finger-pointing in the auto repair world. ASE feels that they can determine your ability to understand an operation without negative inferences.

Answers to the Test Questions for the Additional Test Questions Section 6

1.	B	23.	D	45.	D	67.	B
2.	A	24.	C	46.	A	68.	C
3.	C	25.	D	47.	D	69.	B
4.	A	26.	A	48.	C	70.	C
5.	D	27.	C	49.	C	71.	B
6.	A	28.	C	50.	A	72.	D
7.	C	29.	C	51.	C	73.	C
8.	C	30.	B	52.	B	74.	B
9.	C	31.	D	53.	B	75.	A
10.	D	32.	C	54.	C	76.	C
11.	B	33.	B	55.	C	77.	B
12.	B	34.	C	56.	A	78.	D
13.	A	35.	D	57.	C	79.	A
14.	D	36.	A	58.	D	80.	B
15.	D	37.	B	59.	A	81.	D
16.	D	38.	A	60.	C	82.	B
17.	C	39.	B	61.	C	83.	C
18.	D	40.	B	62.	A	84.	A
19.	A	41.	B	63.	B	85.	D
20.	B	42.	A	64.	B	86.	A
21.	C	43.	C	65.	C	87.	D
22.	D	44.	A	66.	B		

Explanations to the Answers for the Additional Test Questions Section 6

Question #1
Answer A is wrong. Valve stem height is usually measured between the spring seat and the tip of the valve.
Answer B is correct.
Answer C is wrong. Only Technician B is right.
Answer D is wrong. Technician B is right.

Question #2
Answer A is correct. A PCV valve stuck in the open position would not cause blue exhaust smoke.
Answer B is wrong. Worn turbocharger seals on the intake side of the turbo can introduce oil into the intake stream, which enters the combustion chambers and produces blue smoke.
Answer C is wrong. Worn valve seals do not control oil to the valve guides resulting in blue oil smoke.
Answer D is wrong. Worn piston rings can cause oil to enter the combustion chambers either due to poor compression ring seal or poor performance of the oil ring due to deposits between the oil ring scrapers.

Question #3
Answer A is wrong.
Answer B is wrong.
Answer C is correct. It is especially important to thoroughly clean engine oil passages after bearing failure. It is also important that other parts of the lubrication system, which may contain metal particles, be replaced or cleaned to prevent future engine damage.
Answer D is wrong.

Question #4
Answer A is correct.
Answer B is wrong. A broken timing chain could cause a bent pushrod.
Answer C is wrong. A sticking valve could cause a bent pushrod.
Answer D is wrong. Improper valve adjustment could cause a bent pushrod.

Question #5
Answer A is wrong. RTV sealant is never used on threaded fasteners.
Answer B is wrong. Anaerobic sealant fumes will not harm an O2 sensor.
Answer C is wrong. Both Technicians A and B are wrong.
Answer D is correct.

Question #6
Answer A is correct.
Answer B is wrong. Plastigage cannot effectively measure bearing alignment.
Answer C is wrong. A dial indicator could not measure bearing alignment.
Answer D is wrong. A telescoping gauge is not used to measure bearing alignment.

Question #7
Answer A is wrong.
Answer B is wrong.
Answer C is correct. Balance shafts should be checked for runout following the same procedure used to check camshafts for runout. Their journals should be checked for taper following the same procedure used to check crankshaft journals for taper.
Answer D is wrong.

Question #8
Answer A is wrong.
Answer B is wrong.
Answer C is correct. When water pump bearings fail it is usually due to coolant entering the bearing. When the sealed, self-lubricating bearings begin to pit from overheating they will make a growling noise. In some cases the water pump impeller has come off of the pump shaft or the bearing has failed allowing the impeller wheel to drag against the pump housing or timing cover resulting in a loud grinding noise. Answer D is wrong.

Question #9
Answer A is wrong.
Answer B is wrong.
Answer C is correct. Both a bad ground in the cooling fan circuit and a bad wire to the fan relay could prevent the cooling fan from operating.
Answer D is wrong.

Question #10
Answer A is wrong. The cylinder head must be replaced.
Answer B is wrong. When the clearance is excessive, the cylinder head must be replaced.
Answer C is wrong. Inserting bushings is not recommended. The cylinder head must be replaced.
Answer D is correct.

Question #11
Answer A is wrong. Ring grooves should be cleaned using a ring groove cleaning tool.
Answer B is correct.
Answer C is wrong. Only Technician B is right.
Answer D is wrong. Technician B is right.

Question #12
Answer A is wrong. Installing a higher rated thermostat will not cause the engine to warm up faster. It will, however, cause it to operate at a higher temperature.
Answer B is correct.
Answer C is wrong. Only Technician B is right.
Answer D is wrong. Technician B is right.

Question #13
Answer A is correct.
Answer B is wrong. Engine coolant may boil due to low pressure in the system.
Answer C is wrong. Coolant may overflow through the damaged seal or gasket.
Answer D is wrong. If the pressure gets too low or enough coolant is lost, the engine may overheat.

Question #14
Answer A is wrong. This is not a three-angle valve job.
Answer B is wrong. This is not poor contact. The contact shown is correct.
Answer C is wrong. Both Technicians A and B are wrong.
Answer D is correct.

Question #15
Answer A is wrong. Low oil pressure would result in a continuous noise.
Answer B is wrong. Low oil level would result in a continuous noise.
Answer C is wrong. A worn lifter bottom would result in continuous noise.
Answer D is correct.

Question #16
Answer A is wrong. A low reading on two adjacent cylinders may indicate a blown head gasket.
Answer B is wrong. Carbon buildup would cause a high reading.
Answer C is wrong. A low reading on two adjacent cylinders may indicate a cracked cylinder head.
Answer D is correct.

Question #17
Answer A is wrong.
Answer B is wrong.
Answer C is correct. An overtensioned V-belt can damage an alternator front bearing. It can also cause the upper half of the crankshaft front main bearing to wear prematurely.
Answer D is wrong.

Question #18
Answer A is wrong. The cylinder head should not be used as a measurement location.
Answer B is wrong. The measurement should be to the top of the spring seat, not the top shim.
Answer C is wrong. The bottom of the shim should not be used as a measurement location.
Answer D is correct.

Question #19
Answer A is correct. The technician would not run a bottoming tap through the oil gallery bores because they have tapered pipe threads.
Answer B is wrong.
Answer C is wrong.
Answer D is wrong.

Question #20
Answer A is wrong. If the installed height is incorrect, a valve seat may indeed repair the condition, but the question relates to spring seat pressure.
Answer B is correct. Spring shims may be used to correct minor seat pressure issues. In most cases the safer and longer-lived approach is to simply replace the spring.
Answer C is wrong. This is effectively the same answer as A. You should be checking the installed height of the valve stem and the spring during head assembly.
Answer D is wrong.

Question #21
Answer A is wrong. A dye penetrant must be used on aluminum heads.
Answer B is wrong. An electromagnetic-type tester is not used on pistons.
Answer C is correct.
Answer D is wrong. Aluminum intake manifold must be checked using a dye penetrant.

Question #22
Answer A is wrong. Shims are not added to adjust lash.
Answer B is wrong. Shims are not added to adjust lash.
Answer C is wrong. Valve adjustment is required.
Answer D is correct.

Question #23
Answer A is wrong. Excessive taper may require crankshaft grinding.
Answer B is wrong. Out-of-round journals may require crankshaft grinding.
Answer C is wrong. Excessive journal scoring may be removed through crankshaft grinding.
Answer D is correct.

Question #24
Answer A is wrong.
Answer B is wrong.
Answer C is correct. Both the valve spring and the valve stem seal can be replaced without removing the cylinder head from the engine.
Answer D is wrong.

Question #25
Answer A is wrong. Disabling the faulty cylinder will cause engine rpm to drop less than for the other cylinders.
Answer B is wrong. Disabling the faulty cylinder will not cause the engine to stall.
Answer C is wrong. Both Technicians A and B are wrong.
Answer D is correct.

Question #26
Answer A is correct. When the lifters become cupped, it is because they failed to rotate on the camshaft. This can be due to lubrication problems or wear on the cam surface. Once this type of wear pattern has begun, both the camshaft and the lifter must be replaced as a set.
Answer B is wrong. Rocker arms are generally unaffected when the cam or lifters "go flat"
Answer C is wrong.
Answer D is wrong.

Question #27
Answer A is wrong. Cam timing is not adjusted with a ruler.
Answer B is wrong. TDC is not located in this manner.
Answer C is correct.
Answer D is wrong. Valve lash is not adjusted at the timing chain.

Question #28
Answer A is wrong. Spring tension is not measured in this manner.
Answer B is wrong. Spring free height is not measured in this manner.
Answer C is correct. The block deck is checked for warpage using a straightedge and a feeler gauge. Also, minor nicks and burrs can be removed from the deck using a whetstone or a file.
Answer D is wrong.

Question #29
Answer A is wrong. Valve guide height is measured on the other side of the head.
Answer B is wrong. Valve seat depth is not usually measured. Valve stem height is used to determine if there will be a valve installed height issue.
Answer C is correct. The head is being checked with a straight edge and a feeler gauge. If the clearance found between the head and the straight edge is found to be over the manufacturer's specs, the head must be machined or replaced.
Answer D is wrong. Installed height is measured on the other side of the head.

Question #30
Answer A is wrong. A loose timing chain would not affect just two cylinders.
Answer B is correct.
Answer C is wrong. Only Technician B is right.
Answer D is wrong. Technician B is right.

Question #31
Answer A is wrong. Retarded timing would not result in gauge fluctuation.
Answer B is wrong. Advanced timing would not result in gauge fluctuation.
Answer C is wrong. A stuck EGR valve would not result in gauge fluctuation.
Answer D is correct.

Question #32
Answer A is wrong. Residue at the water pump drain hole may indicate a damaged seal or bearings.
Answer B is wrong. Coolant leaking from the water pump is a sign of failure.
Answer C is correct.
Answer D is wrong. A pressure tester should indicate a water pump leak.

Question #33
Answer A is wrong. If the switch is stuck closed, the fan motor will continue to run because the relay receives voltage directly from the battery.
Answer B is correct.
Answer C is wrong. Only Technician B is right.
Answer D is wrong. Technician B is right.

Question #34
Answer A is wrong.
Answer B is wrong.
Answer C is correct. Torque-to-yield head bolts do stretch permanently when they are tightened, and some torque-to-yield bolts cannot be reused.
Answer D is wrong.

Question #35
Answer A is wrong. Stretched main bearing bores cannot be corrected by filing the caps.
Answer B is wrong. Stretched main bearing bores cannot be corrected by replacing the caps.
Answer C is wrong. Both Technicians A and B are wrong.
Answer D is correct.

Question #36
Answer A is correct. The center-to-center measurement is usually only used to determine if the rod is the correct one for the application.
Answer B is wrong. Rod straightness can and should be measured during inspection.
Answer C is wrong. If the small end of the rod is damaged in a full-floating application, the engine rebuild will almost certainly result in disaster. In a press-fit application damage is far less likely, but the rod should be checked anyway.
Answer D is wrong. This is probably the most critical measurement and the part of the rod that is reconditioned. The large end must be perfectly round and the correct size to provide proper bearing crush during assembly.

Question #37
Answer A is wrong.
Answer B is correct. The second measurement should be taken just below the ring ridge to determine the taper of the cylinder. Taper wear occurs to some degree as a normal wear pattern but can be more pronounced in an engine that has been run too cool. This can be due to a thermostat that was stuck in an open position. Taper wear occurs because the top of the cylinder is where the hottest part of the combustion process happens. The bottom of the cylinder is cooler, which creates different clearances in the block, wearing the bottom more than the top.
Answer C is wrong.
Answer D is wrong.

Question #38
Answer A is correct.
Answer B is wrong. The oil rings should be installed first, then the second compression ring, and then the top compression ring.
Answer C is wrong. Only Technician A is right.
Answer D is wrong. Technician A is right.

Question #39
Answer A is wrong. A 0.250 inch (6.35 mm) lift cam and 1.5:1 rocker arms will cause the valve to open 0.375 inch (9.53 mm).
Answer B is correct.
Answer C is wrong. The rocker arm ratio does not determine which type of lifters must be used.
Answer D is wrong. The rocker arm ratio does not determine which type of lifters must be used.

Question #40
Answer A is wrong. A concentricity tester is a useful tool for measuring valve face and seat concentricity.
Answer B is correct.
Answer C is wrong. Blue dye can indicate if a valve has concentricity.
Answer D is wrong. Dial indicators on testers can check concentricity of the valve face and seat.

Question #41
Answer A is wrong.
Answer B is correct. Hammering the dished areas flat again is the best way to solve this common problem.
Answer C is wrong.
Answer D is wrong.

Question #42
Answer A is correct.
Answer B is wrong. Valves having rounded or uneven stem shoulders are replaced, not machined.
Answer C is wrong. Only Technician A is right.
Answer D is wrong. Technician A is right.

Question #43
Answer A is wrong.
Answer B is wrong.
Answer C is correct. Most manufacturers recommend that piston diameter be measured about 3⁄4 inch below the centerline of the wrist pin bore.
Answer D is wrong.

Question #44
Answer A is correct.
Answer B is wrong. Vibration damper rubber should be inspected for looseness.
Answer C is wrong. Vibration damper rubber should be inspected for cracking.
Answer D is wrong. Vibration damper rubber should be inspected for oil soaking.

Question #45
Answer A is wrong. A ridge reamer should be used to remove the ridge.
Answer B is wrong. Sandpaper should not be used to remove the ring ridge.
Answer C is wrong. Both Technicians A and B are wrong.
Answer D is correct.

Question #46
Answer A is correct. When a rod has stretched, it will cause the big end of the rod to become egg shaped, causing wear on the bearings at the parting lines or perpendicular to the direction of piston travel.
Answer B is wrong. Rod twist will generally not show up in bearing wear since the piston or rod will usually fail before the bearing does due to the unusual side loads it creates on both components.
Answer C is wrong. If a bent rod does not cause the same failure as in answer B, it will cause the rod bearing to wear heavily on opposite outside edges at the top and bottom of the bearing halves.
Answer D is wrong. A loose wrist pin will cause the top bearing half to be pounded thinner resulting in bearing failure.

Question #47
Answer A is wrong. The crankshaft needs to be at BDC.
Answer B is wrong. Boots must be installed on the rod bolts to prevent damage to the crankshaft.
Answer C is wrong. Piston rings must be installed correct side up.
Answer D is correct.

Question #48
Answer A is wrong.
Answer B is wrong.
Answer C is correct. Excessive crank endplay can cause a clunk noise when accelerating from a stop. Inadequate endplay can cause damage to the thrust surfaces of the thrust bearings.
Answer D is wrong.

Question #49
Answer A is wrong.
Answer B is wrong.
Answer C is correct. Most timing belt installation procedures call for the crankshaft to be rotated until piston number 1 is at TDC.
Answer D is wrong.

Question #50
Answer A is correct.
Answer B is wrong. Engine stalling may be caused by a PCV valve stuck in the open position.
Answer C is wrong. Rough idle may occur if a PCV valve is stuck in the open position.
Answer D is wrong. A lean air/fuel mixture will result from a PCV valve stuck in the open position.

Question #51
Answer A is wrong.
Answer B is wrong.
Answer C is correct. Excessive cam bearing clearance or a grounded warning indicator circuit could cause the oil pressure light to remain on while the engine is running.
Answer D is wrong.

Question #52
Answer A is wrong.
Answer B is correct. The technician is unlikely to rotate the adjusting nut CW two turns at a time since this could cause the valve and piston to collide, damaging valve train components.
Answer C is wrong.
Answer D is wrong.

Question #53
Answer A is wrong. A normal function of a hydraulic lifter is to pre-tension the valve train with oil pressure. Some of the oil pressure is lost when the engine is shut-off particularly on the cylinders with valves open. This is called bleed-down or leak-down. Excessive bleed-down will cause valve train noise.
Answer B is correct. Lifters are somewhat convex on the bottom if they are good.
Answer C is wrong. Valve lifters should be replaced if the bottom is pitted.
Answer D is wrong. Valve lifters should be replaced if the bottom is flat.

Question #54
Answer A is wrong.
Answer B is wrong.
Answer C is correct. Both a plugged transfer hose and a cracked filler neck soldered joint could prevent coolant from returning to the radiator when the engine cools.
Answer D is wrong.

Question #55
Answer A is wrong. Worn piston rings are likely to cause this condition.
Answer B is wrong. An obstructed PCV vacuum hose is likely to cause this condition.
Answer C is correct.
Answer D is wrong. A clogged PCV valve is likely to cause this condition.

Question #56
Answer A is correct. Obviously the LEAST-Likely cause is a cracked block.
Answer B is wrong. If valve timing is off, long cranks or no-start may be the result.
Answer C is wrong. In a fuel-injected application, a faulty pump could cause long crank time. In a carbureted application it is more likely that the engine would start and die since fuel is held in the carburetor float bowl.
Answer D is wrong. A stuck open EGR valve could cause long crank times due to leaning out fuel mixture on carbureted engines. Fuel-injected applications would likely over rev on initial start up or start with very low idle control motor position.

Question #57
Answer A is wrong.
Answer B is wrong.
Answer C is correct. On an engine equipped with electronic fuel injection, a loose intake manifold may cause both engine noise and poor vehicle driveability.
Answer D is wrong.

Question #58
Answer A is wrong. A belt tension gauge is used to check tension on standard V-belts.
Answer B is wrong. Measuring belt deflection on a V-ribbed belt is not the best method.
Answer C is wrong. A squealing noise at idle would indicate a loose or worn belt.
Answer D is correct.

Question #59
Answer A is correct. Some TTY bolts are reusable.
Answer B is wrong.
Answer C is wrong.
Answer D is wrong. Many newer engines are using TTY bolts for connecting rods, oil pans, main bearings, and intake manifolds.

Question #60
Answer A is wrong.
Answer B is wrong.
Answer C is correct. Two common engine oil cooler mounting locations are: inside a radiator tank and ahead of the radiator support.
Answer D is wrong.

Question #61
Answer A is wrong. An anti-collapse spring is found in the lower, not the upper radiator hose.
Answer B is wrong. Low coolant level will not cause this noise.
Answer C is correct.
Answer D is wrong. Contact with power steering fluid causes a hose to become soft and gummy.

Question #62
Answer A is correct.
Answer B is wrong. A ruler would not be used to take the measurement.
Answer C is wrong. Dial calipers are not the correct tool to measure the Plastigage width.
Answer D is wrong. A micrometer could not be used for this measurement.

Question #63
Answer A is wrong.
Answer B is correct. Since we are measuring for clearance all the way around the valve stem, we must take the total side-to-side reading and divide it in half. This is the standard procedure even though it is not common to any other cylindrical measurements.
Answer C is wrong.
Answer D is wrong.

Question #64
Answer A is wrong. A faulty crank sensor would not cause a single cylinder to not contribute in a power balance test.
Answer B is correct. In a batch fire and sequential fuel injection system, a faulty injector could kill the cylinder.
Answer C is wrong. This type of leak would affect all cylinders.
Answer D is wrong. The fuel from the vapor canister would affect all cylinders.

Question #65
Answer A is wrong.
Answer B is wrong.
Answer C is correct. Valve rotators should be cleaned and inspected to be sure that they will move. The rotators allow the valve to turn, which promotes even wear of the valve faces and seats. If a rotator feels sticky or stiff it will not function in the engine.
Answer D is wrong.

Question #66
Answer A is wrong. A press-fit harmonic balancer should be removed using a special balancer remove/install tool.
Answer B is correct.
Answer C is wrong. Only Technician B is right.
Answer D is wrong. Technician B is right.

Question #67
Answer A is wrong. The measurement gives no indication of journal condition.
Answer B is correct.
Answer C is wrong. The measurement would not indicate the camshaft lift.
Answer D is wrong. The measurement gives no indication of bearing clearance.

Question #68
Answer A is wrong. A lean fuel mixture would not cause a sulfur smell.
Answer B is wrong. Coolant leaking into the combustion chambers would cause a gray exhaust color.
Answer C is correct.
Answer D is wrong. A vacuum leak would cause a rough idle that would decrease as engine speed increases.

Question #69
Answer A is wrong. Burned or leaking valves cause a fluctuation between 12 and 18 in. Hg (62 and 41 kPa absolute).
Answer B is correct.
Answer C is wrong. Weak valve springs cause a fluctuation between 10 and 25 in. Hg (69 and 17 kPa absolute).
Answer D is wrong. A leaking head gasket would cause a fluctuation between 7 and 20 in. Hg (79 and 35 kPa absolute).

Question #70
Answer A is wrong. If both exhaust valves were burned, the cylinders would stay at TDC during the test.
Answer B is wrong. If the cylinders had holes in the pistons, they would stay at TDC during the test.
Answer C is correct. A damaged head gasket could allow two neighboring cylinders to leak between them.
Answer D is wrong. Collapsed lifters or adjusters would not cause high cylinder leakage.

Question #71
Answer A is wrong. This tool cannot check combustion chamber cubic centimeters (cc).
Answer B is correct.
Answer C is wrong. Only Technician B is right.
Answer D is wrong. Technician B is right.

Question #72
Answer A is wrong. Worn intake valves would cause air leaks at the throttle body or carburetor.
Answer B is wrong. Worn exhaust valves would cause air leaks at the tailpipe.
Answer C is wrong. A broken PCV valve would not cause air to leak.
Answer D is correct.

Question #73
Answer A is wrong.
Answer B is wrong. It would not be possible to check pressure at the tensioner feed while the engine is running.
Answer C is correct. Most manufacturers that use oil fed tensioners give a spec for the length of the actuator.
Answer D is wrong. The tensioner holds the chain tension so this method would not work.

Question #74
Answer A is wrong. On OHC engines, the timing belt or chain does not have to be removed from the block before the head can be removed. In many cases, the camshaft sprocket is disconnected from the cam and the cylinder head is lifted from the engine, leaving the belt or chain in place.
Answer B is correct.
Answer C is wrong. Only Technician B is right.
Answer D is wrong. Technician B is right.

Question #75
Answer A is correct. Most batteries will read a static voltage of 12 to 12.7 volts. You start here to verify that there is enough power to turn the starter over.
Answer B is wrong. Spark plugs might be removed if the engine appears to have a hydro-lock condition due to coolant or fuel in the cylinders.
Answer C is wrong. Spark has no relevance to cranking.
Answer D is wrong. The starter solenoid may be activated by a starter button if it is suspected that the ignition switch or neutral safety switch is at fault, but the solenoid should never be bypassed with a starter button. The starter button could not carry the current.

Question #76
Answer A is wrong.
Answer B is wrong.
Answer C is correct. Both stuck piston rings and a plugged oil drain passage in the cylinder head may allow excessive oil to enter the cylinders. This oil would produce blue/gray smoke when it burned.
Answer D is wrong.

Question #77
Answer A is wrong. The seals are installed after the valves are installed.
Answer B is correct.
Answer C is wrong. Only Technician B is right.
Answer D is wrong. Technician B is right.

Question #78
Answer A is wrong. The thread created will be the same size as the original thread.
Answer B is wrong. The thread created will not be smaller than the original thread.
Answer C is wrong. The thread created will not be larger than the original thread.
Answer D is correct.

Question #79
Answer A is correct.
Answer B is wrong. A clogged radiator would not cause this.
Answer C is wrong. Only Technician A is right.
Answer D is wrong. Technician A is right.

Question #80
Answer A is wrong. Coating both sides of the rubber seals increases the likelihood that the seals will be squeezed out of place when the intake manifold bolts are tightened.
Answer B is correct.
Answer C is wrong. Only Technician B is right.
Answer D is wrong. Technician B is right.

Question #81
Answer A is wrong. Excessive runout would cause binding.
Answer B is wrong. Improperly installed bearings would cause binding.
Answer C is wrong. Bore misalignment would cause binding.
Answer D is correct.

Question #82
Answer A is wrong. Fluorescent light will not cause the dye to glow.
Answer B is correct.
Answer C is wrong. A strobe light will not cause the dye to glow.
Answer D is wrong. Infrared light will not cause the dye to glow.

Question #83
Answer A is wrong. The technician should verify that the woodruff key is in place.
Answer B is wrong. The technician should verify that the oil slinger is in place.
Answer C is correct.
Answer D is wrong. The technician should verify that the oil seal has been lubricated.

Question #84
Answer A is correct.
Answer B is wrong. If the starter ring gear on an automatic transmission flywheel is damaged, the flywheel/ring gear assembly must usually be replaced.
Answer C is wrong. Only Technician A is right.
Answer D is wrong. Technician A is right.

Question # 85
Answer A is wrong. Manifold bolt torque will not cause an isolated crack, such as in the exhaust crossover, in a V-type engine.
Answer B is wrong. Torque sequence will not cause this type of problem either.
Answer C is wrong. If the EGR passage is restricted with carbon, it could lead to higher combustion temperatures but would not generate enough heat to cause the manifold to crack.
Answer D is correct. A stuck shut heat riser valve will force exhaust gases through the underside of the intake manifold at all times, resulting in manifold overheating and possibly cracking.

Question #86
Answer A is correct.
Answer B is wrong. Camshaft runout cannot be measured with the camshaft still mounted in the engine.
Answer C is wrong. Only Technician A is right.
Answer D is wrong. Technician A is right.

Question #87
Answer A is wrong. Using a screwdriver to pierce and pull on the hose may damage the heater core.
Answer B is wrong. Use of a pliers and excessive force may damage the heater core.
Answer C is wrong. Use of a prying tool and excessive force may damage the heater core.
Answer D is correct.

Glossary

Air conditioning The process of adjusting or regulating, by heating or cooling, the quality, temperature, and humidity of air.

Aluminum A nonferrous metal that is light in weight yet can be stronger than steel when mixed with the proper alloys. It is easily cast and machined.

Balance shaft A shaft with counterweights designed to prevent vibration of rotating parts.

Battery BatteryA device for storing electrical energy in chemical form.

Belt A device used to drive the water pump and other accessory power-driven devices.

Block deck The flat surface of the main casting of an engine on which the head attaches.

Cam An abbreviation for camshaft; a device having lobes, driven by the crankshaft via gears, a chain, or a belt that opens and closes the intake and exhaust valves.

Cam follower A term used for valve lifter; a hydraulic or mechanical device, in the valve train, that rides on the camshaft lobe to lift the valve off its seat.

Camshaft journal That part of the camshaft that turns in a bearing.

Cast iron A term used for various cast ferrous alloys containing at least 2% carbon. It is used for many different parts on vehicles.

Catalytic converter An automotive exhaust system component, made of stainless steel, containing a catalyst that reduces hydrocarbons, carbon monoxide, and nitrogen oxides present in the engine exhaust gases.

Coil A term used to describe a spring or an electrical device using many turns (coils) of wire such as springs, ignition coils, solenoids, and relays.

Connecting-rod bearing The bearing of a connecting rod that rotates on the crankshaft.

Cooling system The system that circulates coolant through the engine to dissipate its heat.

Counterbalance shafts One or more rotating shafts found on some engines to counteract the natural vibrations of other rotating parts, such as the crankshaft, in that engine. This balance, or counterbalance, shaft usually turns at twice the crankshaft speed and must be timed properly.

Crankshaft The revolving part of a unit that has the function of delivering power or work from the reciprocating motion. Engines have crankshafts, as do air conditioning compressors and air compressors.

Crankshaft sensor An electronic device used to send crankshaft rotating information to the computer.

Crossfiring A condition whereby spark plugs fire out of turn, usually caused by poor spark plug wire insulation.

Cylinder head That part of the engine that covers the cylinders and pistons.

Cylinder wall A term used for cylinder bore; the inside diameter of a cylinder.

Distributor A device used on many engines to direct high-voltage electrical energy from the coil to the spark plugs.

Elongated Not round; egg-shaped.

End play A term used to describe a spacing or clearance involved with a moving part. Crankshafts and camshafts require end play measurements and must be set to manufacturer's specifications during engine assembly.

Engine oil A lubricant formulated for use in an engine.

Engine oil cooler A device used on some high performance engines, police packages, taxis, trucks, turbo-equipped engines, and diesel engines to prevent the engine oil from overheating. These work by using a heat exchanger exposed to air flow or engine coolant.

Exhaust conditioning The burned and unburned gases that remain after combustion.

Exhaust pipe A pipe that connects the exhaust manifold to the muffler or catalytic converter. It is made of heavier material than tailpipes, and sometimes is double layered.

Face-to-seat As in valve face-to-valve seat contact, this refers to the actual sealing area of the valve and seat. It must be the size and angle specified by the engine manufacturer to properly seal the combustion chamber and have a long service life.

Firing order The order in which the engine cylinders fire and deliver power.

Flywheel A round, heavy metal plate attached to the crankshaft of an engine that helps smooth out power strokes and gives the rotating crankshaft momentum to smoothly get to the next power stroke. The clutch and pressure plate help the flywheel transmit power to the drive wheels.

Frozen A mechanical problem developed from a lack of oil or broken internal parts that prevents motion, such as an engine rotating.

Garter spring A small spring placed behind the lip of a lip seal to maintain contact with the rotating part. Most often associated with oil seals.

Gap A space between two adjacent parts or surfaces.

Head That part of an engine that covers the top of the cylinders and pistons.

Heat shield Devices used many places on today's vehicles. One such place is between the starter solenoid and the heat of the engine and exhaust manifold. Another is between electrical wiring or spark plug wiring and any high heat source. Heat shields are also used between catalytic converters and the passenger compartment of the vehicle.

Hone To use abrasive materials to remove material from a surface, or just to change the surface smoothness so the parts involved will work better.

Idler pulley A pulley that is used to adjust the belts on a belt-driven system.

Input A term used for the signals sent to different electronic modules about operating conditions of systems involved.

Intake ductwork All of the connecting ductwork from the throttle body out involved with getting air into the engine.

Jumped As in jumped timing chain or timing belt, resulting in a condition where the valve timing is no longer correct, and the engine will run poorly or might not start.

Keepers Key-like tapered metal locking devices used to hold valve retainers in place.

Lash The clearance between two parts.

Lash adjuster A device much like hydraulic lifters that is usually found on overhead cam engines. Its job is to maintain zero lash between the cam follower and the tip of the valve.

Line boring A machining process that ensures multiple holes that are bored in a cylinder head or block are in line or true. This allows the camshaft or crankshaft installed in these locations to turn freely and function properly. As in line bore alignment of camshaft bearing or main bearing bores, a special boring bar is used that removes metal from all cam or main bearing bores at the same time and in a straight and true line.

Main bearing cap The structural device that holds the crankshaft in place in an engine block.

Muffler A device in the exhaust system used to reduce noise.

No-crank A condition like a frozen engine, a defective starter, or an electrical problem that prevents the engine from rotating when normal attempts are made to start the engine.

No-start A condition where the engine turns over normally but does not start. This could be caused by mechanical problems, fuel system problems, electrical problems, or electronic engine control problems.

Oil cooler A heat exchanger used to cool transmission or engine oil. Also may be used to cool power steering or other fluids.

Oil filter A device used to remove impurities, such as abrasive particles, from oil.

Oil pan A removable part of the engine assembly that contains the oil supply.

Out of square As in a valve spring that is not within specification when checked vertically on a flat surface and measured at a 90-degree angle. If installed in an engine, such a spring would cause side pressures on the valve and damage the valve and valve guide.

Overhead cam A camshaft that is mounted in the cylinder head.

Oxygen sensor An electronic device found in the exhaust system that measures the amount of oxygen in the exhaust stream.

Piston pin A precision ground pin, usually hollow, used to attach the connecting rod to the piston. The piston pin can be held in place by a press fit, snaprings, or bolts.

Pressure cap A cap placed on the radiator to allow regulated, above-atmospheric pressure in the cooling system.

Primed Ready; prepared.

Pulley A wheel-shaped device used in a belt-drive system to drive accessory equipment.

Ram air Air forced through the radiator, condenser, and across the engine by the forward movement of the vehicle.

Reluctor A gear-like part of an electronic ignition system. It could have the same number of teeth (or gaps) as cylinders of the engine, or it could have half the number of teeth as the engine has cylinders. As a tooth (or gap) passes by a pickup coil or crankshaft sensor, the

magnetic field changes, and a trigger signal is sent to an electronic control module.

Reluctor ring A gear-like part of the electronic ignition system.

Rotators As in valve rotators-devices that cause the valves in the cylinder head to rotate slightly each time the valve closes. This helps keep the valve seat and valve face clean. Because a different part of the valve face contacts the valve seat each time it closes, the valve also operates at a cooler temperature. This greatly extends valve and seat life.

Rotor A part of the ignition distributor that rotates inside the cap and transfers ignition coil secondary electrical energy from the center tower to the individual spark plug wires.

RTV An abbreviation for Room Temperature Vulcanizing; the trade name for a rubber-like sealing compound.

Select fit As in the main bearings of an engine. Many of today's engines use select fit main bearings. This means the main bearings are no longer all one standard size, or all 0.010 inch (0.254 mm) undersized. The manufacturer mixes and matches bearing halves of small increments of as little as 0.003 inch (0.076 mm) to ensure better oil and noise control at the crank-shaft area.

Selective thrust washer A washer or spacer that is furnished in different sizes to facilitate end play and preload settings.

Skirt A term used to describe the lower part of an engine piston. The skirt contacts the cylinder wall, helps the piston travel in a straight line, and prevents piston slap.

Solenoid An electromechanical device used for a push-pull operation.

Span A term used for the length or space between two parts; as in the span between the air conditioner compressor belt pulley and the alternator pulley is too great. This condition can cause the belts to fly off at high engine speeds or during rapid acceleration.

Spark plug A component of the ignition system that delivers the high-voltage spark to the combustion chamber.

Starter drive The part of the starter motor that engages the ring gear or the flywheel, flex plate, or torque converter, and rotates the engine on start-up.

Starter motor The small electric motor that is used to crank (start) an engine.

Starter solenoid That part that causes the starter drive to engage the flywheel when starting an engine.

Tailpipe The pipe from the muffler or catalytic converter that carries the exhaust gases away from the passenger compartment.

Temperature sensor A term used for various temperature sensing switches or variable resistors on today's vehicles. They can be used to turn cooling fans on and off, to operate dash gauges, to control air conditioners, and to furnish inputs to engine and transmission control computers.

Tensioner A device used with timing belts and timing chains that maintains a constant pressure on the belt or chain to minimize wear and noise, and to take up slack or lost motion. They may be fixed and require periodic adjustment, or they may be automatically adjusted by spring tension or engine oil pressure. Tensioners are also used on accessory drive belts.

Timing belt The belt through which the crankshaft drives the camshaft(s) in an overhead cam engine.

Torque wrench A specially designed tightening device that indicates the amount of torque being exerted on a fastener to enable threaded parts to be tightened to a specified amount.

Torque-to-yield A term used to describe a common method of tightening fasteners on many of today's engines. The procedure is to tighten the fasteners to a fairly low pounds-feet value, then to turn each fastener (in proper tightening sequence) a specified number of degrees. This process is usually used on head bolts, main bearing bolts, and rod bearing bolts. In most cases, new fasteners are required for every reassembly.

True A term used in the automotive industry to signify a part or system is correct and is within specifications. As in the cylinder head gasket sealing surface is true; i.e., it is not warped, scratched, broken, or otherwise damaged.

Valve float A condition that occurs when the valve spring is not capable of closing the valve quickly enough. This usually happens at higher engine speeds, and is aggravated by the valve spring losing some of its tension. Improper sealing of the combustion chamber will occur and the engine will run poorly.

Valve lifter A hydraulic or mechanical device in the valve train that rides on the camshaft lobe to lift the valve off its seat.

Valve rotator A device that rotates the valve while the engine is running.

Valve spring retainer A device on the valve stem that holds the spring in place.

Valve train The parts making up the valve assembly and its operating mechanism.

Warpage As in cylinder head gasket, sealing area no longer being straight and true. This condition will require machine shop work on the cylinder head or a different cylinder head that is not warped.

Water pump A mechanical device used to circulate coolant through the cooling system.

Witness marks Lines scribed on adjacent surfaces of mating parts, before disassembly, to ensure proper alignment when reassembled.

Notes

Notes

Notes

Notes

Notes